E. Corbacho

Introduction to the RAFU method on approximation

E. Corbacho

Introduction to the RAFU method on approximation

LAP LAMBERT Academic Publishing

Impressum / Imprint

Bibliografische Information der Deutschen Nationalbibliothek: Die Deutsche Nationalbibliothek verzeichnet diese Publikation in der Deutschen Nationalbibliografie; detaillierte bibliografische Daten sind im Internet über http://dnb.d-nb.de abrufbar.

Alle in diesem Buch genannten Marken und Produktnamen unterliegen warenzeichen-, marken- oder patentrechtlichem Schutz bzw. sind Warenzeichen oder eingetragene Warenzeichen der jeweiligen Inhaber. Die Wiedergabe von Marken, Produktnamen, Gebrauchsnamen, Handelsnamen, Warenbezeichnungen u.s.w. in diesem Werk berechtigt auch ohne besondere Kennzeichnung nicht zu der Annahme, dass solche Namen im Sinne der Warenzeichen- und Markenschutzgesetzgebung als frei zu betrachten wären und daher von jedermann benutzt werden dürften.

Bibliographic information published by the Deutsche Nationalbibliothek: The Deutsche Nationalbibliothek lists this publication in the Deutsche Nationalbibliografie; detailed bibliographic data are available in the Internet at http://dnb.d-nb.de.

Any brand names and product names mentioned in this book are subject to trademark, brand or patent protection and are trademarks or registered trademarks of their respective holders. The use of brand names, product names, common names, trade names, product descriptions etc. even without a particular marking in this work is in no way to be construed to mean that such names may be regarded as unrestricted in respect of trademark and brand protection legislation and could thus be used by anyone.

Coverbild / Cover image: www.ingimage.com

Verlag / Publisher:
LAP LAMBERT Academic Publishing
ist ein Imprint der / is a trademark of
OmniScriptum GmbH & Co. KG
Heinrich-Böcking-Str. 6-8, 66121 Saarbrücken, Deutschland / Germany
Email: info@lap-publishing.com

Herstellung: siehe letzte Seite /
Printed at: see last page
ISBN: 978-3-659-79747-7

To my parents Eduardo and María
To my understanding wife Pilar
To my children Eduardo and Alicia

The RAFU method on approximation uses radical functions (RAFU functions) of odd index to approximate an arbitrary function from the knowledge of its values, or its approximate values, at a set of points.

The main goal of this book is to show to the scientific community all the results published until now about this approximation procedure [Co1, Co2, Co3] in order that they could be applied in the cases that the researchers consider appropriate to do it.

As far as we know, the work is original. It does not impress with the difficulties it overcomes. It does not contain complicated calculations or reasonings, but we think that the importance of this technique to solve approximation problems will balance the deficiency of difficulties.

The book is organized as follows. In Chapter 1 we will give some ideas about the begining of the RAFU method. In Chapter 2 we will study the aproximation of continuous functions in different practical cases, we will give uniform approximation algorithms, we will obtain the degree of uniform approximation and we will make a comparative analysis of this degree of approximation. Chapter 3 will be devoted to the topological foundation of the RAFU method. Finally, in Chapter 4 we will prove that the RAFU method can be used to approximate functions of $C_0(\mathbb{R})$ and $C_{00}(\mathbb{R})$, Riemann integrable functions, Lebesgue integrable functions, functions of $L^p[a,b]$ and $L^p(\mathbb{R})$, $1 \leq p < \infty$ and measurable functions. Moreover, Riemann integrals can be approximated by the integrals of the RAFU functions.

We finish this introduction with some acknowledgements. The author is grateful to D. Francisco Lavado Martínez, my math teacher in the high school. One loves mathematics thanks to him. The author would like to thank D. Rufino Rodríguez Sánchez. His helpfull suggestions have done possible the RAFU method appeared. The author also wishes to thank Professor Manuel Mota Medina of the Department of Mathematics of the University of Extremadura for providing a stimulating scientific example.

Contents

Contents   **1**

1   The beginning of the RAFU method   **3**

   1.1   Introduction   3

   1.2   The first idea   4

   1.3   Approximation to a step function   5

       1.3.1   Particular case   5

       1.3.2   General case   7

2   Uniform approximation in $C[a, b]$ **13**

   2.1   Introduction   13

   2.2   A set of RAFU functions uniformly dense in $C[a, b]$ 14

   2.3   Approximation properties of the RAFU functions   15

       2.3.1   Approximation by parts   15

       2.3.2   Influence of the change of functions   18

       2.3.3   Approximation in different practical cases   18

            2.3.3.1   Approximation from average samples   19

            2.3.3.2   Approximation from local average samples   19

            2.3.3.3   Approximation from linear combinations   20

            2.3.3.4   Approximation from approximate values   20

   2.4   Uniform approximation algorithms   23

   2.5   The degree of approximation by RAFU functions   29

2.6   Comparative analysis of the degree of approximation 33

2.7   Uniform approximation from non uniform spaced points 36

3 The topological foundation of the RAFU method   **41**

3.1   Introduction 41

3.2   A RAFU linear space uniformly dense in $C[a, b]$ 42

3.3   The *best* approximation problem 46

4 Approximation in different smoothness spaces with the RAFU method   **49**

4.1   Introduction 49

4.2   Approximation on $C_0(\mathbb{R})$ and $C_{00}(\mathbb{R})$ 50

4.3   Approximation to a Riemman integrable function 50

4.4   Approximation to a Lebesgue integrable function 53

4.5   Approximation on $L^p[a, b]$ and $L^p(\mathbb{R})$, $1 \le p < \infty$ 53

4.6   Approximation to a measurable function 55

Bibliography   **57**

Chapter 1

The beginning of the RAFU method

1.1 Introduction

Definition 1.1 *Let f be an arbitrary function defined in $[a, b]$ and let $a = x_0 < x_1 < \ldots < x_n = b$ be a partition of $[a, b]$ for each natural n. The RAFU method on approximation is an approximation procedure to the function f by a sequence of radical functions $(C_n)_n$ defined by the formula*

$$C_n(x) = f(x_1) + \sum_{p=2}^{n} [f(x_p) - f(x_{p-1})] \cdot F_n(x_{p-1}, x) \tag{1.1}$$

being $F_n(x_p, x) = \dfrac{{}^{2n+1}\!\sqrt{x_p - a} + {}^{2n+1}\!\sqrt{x - x_p}}{{}^{2n+1}\!\sqrt{b - x_p} + {}^{2n+1}\!\sqrt{x_p - a}}$, $p = 1, \ldots, n-1$. We will say that the functions of the type (1.1) are RAFU functions. RAFU approximation will be the approximation procedure based in the use of the RAFU functions.

This Chapter is devoted to describe the beninning of the RAFU method. In Section 2 we will recall the first idea. In Section 3 we will approximate step functions by means of the RAFU functions defined in (1.1). From these results the RAFU method has been

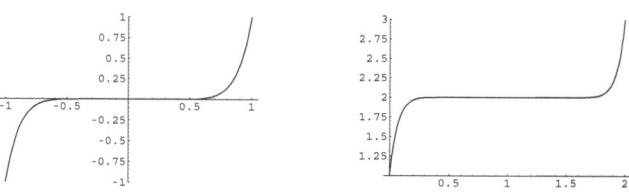

Figure 1.1:

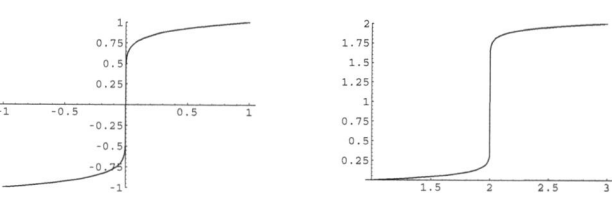

Figure 1.2:

constructed. The reader can see [Co1, Co2, Co3] for details about this approximation procedure.

1.2 The first idea

Consider the functions $f_1(x) = x^9$ defined in $[-1, 1]$ and $f_2(x) = 2 + (x-1)^{15}$ defined in $[0, 2]$ (Fig. (1.1)). Next, observe the corresponding inverse functions $g_1(x) = \sqrt[9]{x}$ defined in $[-1, 1]$ and $g_2(x) = 1 + \sqrt[15]{x - 2}$ defined in $[1, 3]$ (Fig. (1.2)).

If one looks closely, the graphics of these radical functions of odd index are similar to the graphics of certain step functions. Moreover this similarity increases with n. At that moment, many questions appeared. Would be possible to use these functions to approximate an arbitrary step function? And a continuous function? What would be the degree of this approximation? What kinds of functions could we approximate? Is there any topological root for these approximations? Would be easy to work with these

4

radical functions in order to solve these problems? In this way it began the construction of the RAFU method.

1.3   Approximation to a step function

Here, our main goal is *to approximate* an arbitrary step function by means of a family of RAFU functions of the type (1.1) and then to study the sense of this approximation. First and foremost, it is convenient to know along this book that Proposition (1.1) holds.

Proposition 1.1 *It verifies that*

1. $\lim_{n \to +\infty} \sqrt[2n+1]{x} = \begin{cases} -1 & if \quad x < 0 \\ 0 & if \quad x = 0 \\ 1 & if \quad x > 0 \end{cases}$

2. $\lim_{n \to +\infty} \sqrt[2n+1]{n} = 1$

3. $\lim_{n \to +\infty} \sqrt[2n+1]{-n} = -1$

4. (a) $\lim_{n \to +\infty} \sqrt[2n+1]{\frac{1}{n}} = 1$

   (b) $\lim_{n \to +\infty} \sqrt[2n+1]{-\frac{1}{n}} = -1$

The proof is based in elementary calculations.

1.3.1   Particular case

Given the step function

$$E(x) = \begin{cases} k_1 & if \quad x_0 \leq x \leq x_1 \\ k_2 & if \quad x_1 < x \leq x_2 \end{cases} , \qquad k_1, k_2 \in \mathbb{R}$$

we consider the family of radical functions

5

$$c_n(x) = M_n + N_n \cdot \sqrt[2n+1]{x - x_1}, \quad n \in \mathbb{N} \quad M_n, N_n \in \mathbb{R}$$

Solving the following system

$$\begin{cases} k_1 = M_n + N_n \cdot \sqrt[2n+1]{x_0 - x_1} \\ k_2 = M_n + N_n \cdot \sqrt[2n+1]{x_2 - x_1} \end{cases}$$

it follows that

$$N_n = \frac{k_2 - k_1}{\sqrt[2n+1]{x_2 - x_1} + \sqrt[2n+1]{x_1 - x_0}}$$

and

$$M_n = k_1 + \frac{(k_2 - k_1) \cdot \sqrt[2n+1]{x_1 - x_0}}{\sqrt[2n+1]{x_2 - x_1} + \sqrt[2n+1]{x_1 - x_0}}$$

So, from Proposition (1.1), it is easy to check that

$$\lim_{n \to +\infty} c_n(x) = \begin{cases} k_1 & if \quad x_0 \le x < x_1 \\ \frac{k_1 + k_2}{2} & if \quad x = x_1 \\ k_2 & if \quad x_1 < x \le x_2 \end{cases}$$

Example 1.1 *In Figure (1.3) we represent the functions c_n for $n = 9$ and $n = 57$ which approximate to the step function*

$$E(x) = \begin{cases} -10 & if \quad -100 \le x \le 20 \\ 50 & if \quad 20 < x \le 300 \end{cases}$$

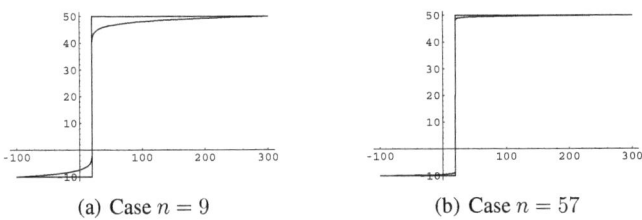

(a) Case $n = 9$ (b) Case $n = 57$

Figure 1.3: Approximation to a step function

1.3.2 General case

Let E_m be the step function

$$E_m(x) = \begin{cases} k_1 & if \quad x_0 \le x \le x_1 \\ k_2 & if \quad x_1 < x \le x_2 \\ \quad ... \\ k_{m-1} & if \quad x_{m-2} < x \le x_{m-1} \\ k_m & if \quad x_{m-1} < x \le x_m \end{cases} \qquad k_1, ..., k_m \in \mathbb{R} \qquad (1.2)$$

Here, we define the functions

$$f_1(x) = k_1 \quad if \quad x_0 \le x \le x_m$$

$$f_p(x) = \begin{cases} 0 & if \quad x_0 \le x \le x_{p-1} \\ k_p - k_{p-1} & if \quad x_{p-1} < x \le x_p \end{cases} \qquad p = 2, ..., m$$

By definitions, it verifies that

$$E_m(x) = \sum_{i=1}^{m} f_i(x) \quad \forall x \in [x_0, x_m]$$

The functions f_i can be approximated by the sequences $(c_{i,n})_n$ defined by the following

7

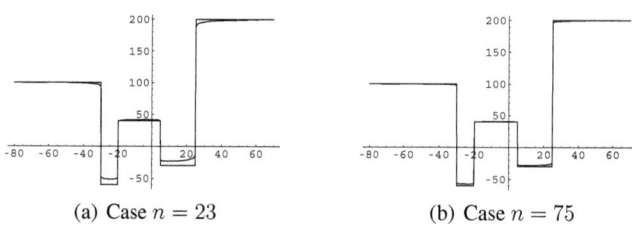

(a) Case $n = 23$ (b) Case $n = 75$

Figure 1.4: Approximation to a step function

formulas

For f_1 we consider $c_{1,n}(x) = k_1, \quad n \in \mathbb{N}$

For each f_p, $p = 2, ..., m$ and $n \in \mathbb{N}$, we consider $c_{p,n}(x) = M_{p,n} + N_{p,n} \cdot \sqrt[2n+1]{x - x_{p-1}}$ where $N_{p,n}$ and $M_{p,n}$ are obtained considering the conditions $c_{p,n}(x_0) = 0$ and $c_{p,n}(x_m) = k_p - k_{p-1}$.

$$N_{p,n} = \frac{k_p - k_{p-1}}{\sqrt[2n+1]{x_m - x_{p-1}} + \sqrt[2n+1]{x_{p-1} - x_0}} \tag{1.3}$$

$$M_{p,n} = \frac{(k_p - k_{p-1}) \cdot \sqrt[2n+1]{x_{p-1} - x_0}}{\sqrt[2n+1]{x_m - x_{p-1}} + \sqrt[2n+1]{x_{p-1} - x_0}} \tag{1.4}$$

Example 1.2 *In Figure (1.4) we represent the functions $C_{5,n} = \sum_{i=1}^{5} c_{i,n}$ for $n = 23$ and $n = 75$ that approach to the step function*

$$f(x) = \begin{cases} 100 & if & -80 \le x \le -30 \\ -60 & if & -30 < x \le -20 \\ 40 & if & -20 < x \le 5 \\ -30 & if & 5 < x \le 25 \\ 200 & if & 25 < x \le 70 \end{cases}$$

Proposition 1.2 *Let $C_{m,n}$ be the function $C_{m,n} = \sum_{i=1}^{m} c_{i,n}(x)$ or, in another form,*

$$C_{m,n}(x) = k_1 + \sum_{p=2}^{m} [k_p - k_{p-1}] \cdot \frac{\sqrt[2n+1]{x_{p-1} - x_0} + \sqrt[2n+1]{x - x_{p-1}}}{\sqrt[2n+1]{x_m - x_{p-1}} + \sqrt[2n+1]{x_{p-1} - x_0}} \qquad (1.5)$$

Then, it follows that

$$\lim_{n \to +\infty} C_{m,n} = \begin{cases} k_1 & if \quad x_0 \leq x < x_1 \\ \frac{k_1+k_2}{2} & if \quad x = x_1 \\ k_2 & if \quad x_1 < x < x_2 \\ \frac{k_2+k_3}{2} & if \quad x = x_2 \\ \quad \cdots \\ \frac{k_{m-1}+k_m}{2} & if \quad x = x_{m-1} \\ k_m & if \quad x_{m-1} < x \leq x_m \end{cases}$$

Proof 1.1 *By the properties of the limits and the definitions of the functions.*

Next, we will focus attention on the convergence of the sequence $(C_{m,n})_n$ to the function E_m defined in (1.2).

Let $\beta > 0$ be such that $(x_i - \beta, \, x_i + \beta) \cap (x_j - \beta, \, x_j + \beta) = \emptyset$ where $i \neq j$ and $i, j \in \{1, \, ..., \, m - 1\}$. Then,

Proposition 1.3 *The limit $\lim_{n \to +\infty} C_{m,n} = f$ is uniform on*

$$[x_0, \, x_1 - \beta] \cup [x_1 + \beta, \, x_2 - \beta] \cup ... \cup [x_{m-1} + \beta, \, x_m]$$

Proposition 1.4 *For all $\varepsilon > 0$, there exists $n_0 \in \mathbb{N}$ such that for $n > n_0$ it follows that*

1. $| \, C_{m,n}(x) - E_m(x) \, | < | \, k_{j+1} - k_j \, | + \varepsilon$

2. $|C_{m,n}(x) - (k_j \cdot (1 - \alpha) + k_{j+1} \cdot \alpha)| < \varepsilon$

where $x \in (x_j - \beta, \, x_j + \beta)$, $\quad j = 1, ..., m - 1$ and $\alpha \in (0, \, 1)$.

9

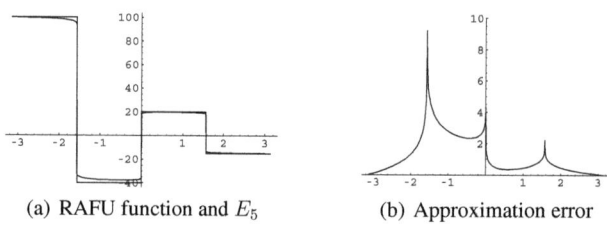

(a) RAFU function and E_5 (b) Approximation error

Figure 1.5: Approximation with the RAFU method

Proofs of Propositions (1.3) and (1.4) can be seen in [Co1].

Until now, we have seen that the RAFU functions can be used to approximate a step function. The other questions mentioned at the end of Section 2 will be answered along the following chapters.

It is not our interest in this work to make a comparison with other approximation procedures but in Example (1.3) we show some differences between the approximation to a step functions by the RAFU functions and the Fourier's polynomial.

Example 1.3 *Let E_5 be the step function defined by*

$$E_5(x) = \begin{cases} 100 & if \quad -\pi \leq x \leq -\frac{\pi}{2} \\ -40 & if \quad -\frac{\pi}{2} < x \leq 0 \\ 20 & if \quad 0 < x \leq \frac{\pi}{2} \\ -15 & if \quad \frac{\pi}{2} < x \leq \pi \end{cases}$$

For $n = 30$, in Figure (1.5) we represent the RAFU approximation to f. In Figure (1.6) we do the same with the Fourier's polynomial.

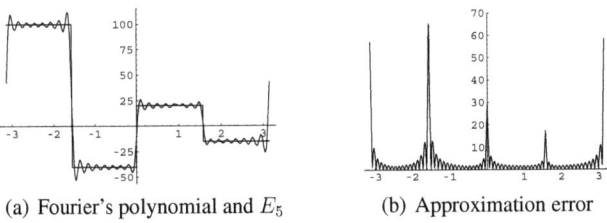

(a) Fourier's polynomial and E_5 (b) Approximation error

Figure 1.6: Fourier's polynomial approximation

Chapter 2

Uniform approximation in $C[a,b]$

2.1 Introduction

In Section 2 we will prove that a family of RAFU functions, $\mathbb{C}$, is uniformly dense in $C[a,b]$. Given $f \in C[a,b]$, the mathematical expression of the terms of the sequence that converges uniformly to f will be determined. The uniform stability of the approximation procedure by radical functions improves the instability of the interpolation by means of polynomials and this is a special contribution of this work.

In Section 3 we will study some approximation properties of the RAFU functions. In the fourth Section, we will propose several useful algorithms to approximate uniformly a continuous function in different practical cases by using the 4.1.0.0 Mathematica program. Some examples will be showed.

The degree of uniform approximation of the RAFU method will be obtained in Section 5. Section 6 will be devoted to compare this degree of approximation with the approximation by arbitrary polynomials and Bernstein polynomials. Finally, we will deal with the uniform approximation from non uniform spaced points in the last section.

2.2 A set of RAFU functions uniformly dense in $C[a, b]$

Definition 2.1 *We will say that a family $\mathcal{F} \subset C[a, b]$ is uniformly dense on $C[a, b]$ if each function $f \in C[a, b]$ is a uniform limit of functions from $\mathcal{F}$.*

Given a step function defined in a closed interval $[a, b]$ as (1.2), by Propositions (1.3) and (1.4), we know that there exists a radical continuous functions sequence uniformly convergent in all the closed interval except in the small open neighbourhoods centered at the points of step $x_1, ..., x_{m-1}$. Moreover, the approximating functions take values from k_i to k_{i+1} in each open neighbourhood centered at x_i, for any $i = 1, ..., m - 1$. So, these results encourage us to prove that the sequences $C_{m,n})_n$ can be employed to approximate an arbitrary continuous function $f \in C[a, b]$.

Consider $f \in C[a, b]$ and let $f(x_0)$, $f(x_1)$, ..., $f(x_m)$ be its values at the points of the partition of $[a, b]$ given by $P = \{x_0 = a, x_1, ..., x_m = b\}$. From these data, we can define step functions, E_m, in $[a, b]$ that approximate to the function f. Even more, since f is uniformly continuous in $[a, b]$, there exist sequences of step functions $(E_m)_m$ such that

$$\lim_{n \to +\infty} E_m = f$$

and these limits are uniform in $[a, b]$. In fact, the proof of Lemma (2.1) is trivial.

Lemma 2.1 *Let f be a continuous function defined in an interval $[a, b]$. Then, there exists a sequence of step functions $(E_m)_m$ defined in $[a, b]$ that converges uniformly to f in $[a, b]$.*

For each natural $n \geq 2$, we will designate by C_n the subset of $C[a, b]$ formed for the functions $C_{m,n}$ with $m = n$. Moreover, the points of the partition $P_n = \{a = x_0, x_1, ..., x_{n-1}, x_n = b\}$ of $[a, b]$ which serve us to define $C_{m,n}$ will be $x_j = a + j \cdot \frac{b-a}{n}$, where $j = 0, 1, ..., n$. In this case, the functions $C_{n,n} = C_n$ from C_n have the form

$$C_n(x) = M_1 + \sum_{i=2}^{n} \left(M_i + N_i \cdot \sqrt[2n+1]{x - x_{i-1}} \right)$$

14

where M_i and N_i are defined as (1.3) and (1.4) respectively. We will also denote by $\mathbb{C}^*$ the set $\mathbb{C}^* = \cup_{n \geq 2} \mathbb{C}_n$. With this notation, the following result holds.

Theorem 2.1 *The family $\mathbb{C}^*$ is uniformly dense on $C\left[a, b\right]$.*

Proof 2.1 *Propositions (1.3) and (1.4) hold when $m = n$ and the proofs are analogous. The remainder is a consequence of Lemma (2.1), of the uniform continuity of a continuous function in a compact interval $\left[a, b\right]$ and the definition of $\mathbb{C}^*$.*

Remark

Besides proving the uniform density of $\mathbb{C}^*$ on $C\left[a, b\right]$, we also know the expression of each term C_n of the sequence $(C_n)_n$ which converges uniformly to f in $\left[a, b\right]$:

$$C_n(x) = f(a) + \sum_{j=2}^{n} [f(x_j) - f(x_{j-1})] \cdot \frac{\sqrt[2n+1]{x_{j-1} - a} + \sqrt[2n+1]{x - x_{j-1}}}{\sqrt[2n+1]{b - x_{j-1}} + \sqrt[2n+1]{x_{j-1} - a}} \quad (2.1)$$

with $x_j = a + j \cdot \frac{b-a}{n}$, $j = 0, 1, ..., n$.

2.3 Approximation properties of the RAFU functions

Next, in the following subsections, we will use the mentioned approximation procedure to answer some typical subjects in Numerical Analysis.

2.3.1 Approximation by parts

Let $a < c < b$ be. By means of this approximation method we can obtain a uniform approximation to f in $\left[a, b\right]$ from uniform approximations to f in $\left[a, c\right]$ and $\left[c, b\right]$

Proposition 2.1 *Let f be a continuous function in $\left[a, b\right]$ and let $(C_n)_n$ be the uniformly convergent sequence to f in $\left[a, b\right]$ defined by (2.1). Let M_1 and M_2 be real numbers such that $\left[a, b\right] \subset \left[M_1, M_2\right]$. Then, from $(C_n)_n$ we can construct a sequence $(C'_n)_n$ defined in $\left[M_1, M_2\right]$ satisfying:*

1. $(C'_n)_n$ converges uniformly to f on $[a + \beta, b - \beta]$, $\forall \beta > 0$.

2. $(C'_n)_n$ converges uniformly to 0 on $[M_1, a - \beta] \cup [b + \beta, M_2]$, $\forall \beta > 0$.

3. For any $\epsilon > 0$, there exists $n_0 \in \mathbb{N}$ such that for each $n \geq n_0$ it follows that:

   (a) $\left| C'_n(x) - f(a) \cdot \alpha \right| < \epsilon$ where $\alpha \in (0, 1)$ y $x \in (a - \beta, a + \beta)$.

   (b) $\left| C'_n(x) - f(b) \cdot \alpha \right| < \epsilon$ where $\alpha \in (0, 1)$ y $x \in (b - \beta, b + \beta)$.

Proof 2.2 *For each partition* $P_n = \{a = x_0, ..., x_n = b\}$ *of* $[a, b]$ *we construct a new partition* $Q_n = \{x'_0 = M_1, x'_1 = a, ..., x'_{n+1} = b, x'_{n+2} = M_2\}$ *of* $[M_1, M_2]$ *and we define the step function*

$$
E'_n(x) = \begin{cases}
0, & x \in [M_1, a) \\
f(a), & x \in [a, x_1] \\
f(x_2), & x \in (x_1, x_2] \\
\quad \cdots \\
f(x_{n-1}), & x \in (x_{n-2}, x_{n-1}] \\
f(b), & x \in (x_{n-1}, b] \\
0, & x \in (b, M_2]
\end{cases}
$$

For each E'_n *we define the function* C'_n *by the formula given in (2.1). Then,* $(C'_n)_n$ *verifies 1 applying Theorem (2.1),* $(C'_n)_n$ *verifies 2 by Proposition (1.3) and* $(C'_n)_n$ *satisfies 3 by Proposition (1.4).*

Remark
It is not necessary to know the values of f in $[M_1, M_2] - [a, b]$.

Corollary 2.1 *Let* f *be a continuous function defined in* $[a, c]$ *and let* $a < b < c$ *be. Consider the sequences* $(C_{n,1})_n$ *defined in* $[a, b]$ *and* $(C_{n,2})_n$ *defined in* $[b, c]$ *by the formula (2.1) which converge uniformly to* f *in* $[a, b]$ *and* $[b, c]$ *respectively. Then,* *from $(C_{n,1})_n$ and $(C_{n,2})_n$ we can construct a new sequence of radical functions $(C_n)_n$ defined in $[a, c]$ uniformly convergent to f in $[a, c]$.*

Proof 2.3 *From the sequences $(C_{n,1})_n$ and $(C_{n,2})_n$ we construct $(C'_{n,1})_n$ and $(C'_{n,2})_n$ defined in $[a, c]$ as in Proposition (2.1).*

Now, for each n, we define the function C_n in the interval $[a, c]$ by the formula

$$C_n(x) = C'_{n,1}(x) + C'_{n,2}(x)$$

Since f is uniformly continuous in $[a, c]$, for each $\epsilon > 0$

$$\exists \delta > 0 : \mid x - x' \mid < \delta \Rightarrow \mid f(x) - f(x') \mid < \frac{\epsilon}{8}$$

From now on we will make partitions of $[a, b]$ and $[b, c]$ into intervals of length $< \delta$.

By Proposition (2.1) 1, 2 it follows that for a fixed $\beta > 0$, there exists $n_1 \in \mathbb{N}$ such that if $n \geq n_1$ then

$$|C_n(x) - f(x)| = \left| C'_{n,1}(x) + C'_{n,2}(x) - f(x) \right| < \epsilon$$

being $x \in [a, b - \beta] \cup [b + \beta, c]$.

For these $\epsilon > 0$ and $\beta > 0$ there exists $n_2 \in \mathbb{N}$ such that if $n \geq n_2$ then

$$\left| C'_{n,1}(x) - f(b) \cdot \theta \right| < \frac{\epsilon}{4}, \quad \theta \in (0, 1), \ x \in (b - \beta, b + \beta)$$

Moreover, for the same $\epsilon > 0$ and $\beta > 0$ there exists $n_3 \in \mathbb{N}$ such that if $n \geq n_3$

$$\left| C'_{n,2}(x) - f(b) \cdot \mu \right| < \frac{\epsilon}{4}, \quad \mu \in (0, 1), \ x \in (b - \beta, b + \beta)$$

Finally, if $n \geq max\{n_2, n_3\}$ and we consider $\alpha \in (0, 1)$

$$|C_n(x) - f(x)| = \left| C'_{n,1}(x) + C'_{n,2}(x) - f(x) \right|$$

$$= \left| C'_{n,1}(x) + C'_{n,2}(x) \pm [f(b) \cdot \theta + f(b) \cdot \mu] - f(x) \right|$$

$$\leq \left| C'_{n,1}(x) - f(b) \cdot \theta \right| + \left| C'_{n,2}(x) - f(b) \cdot \mu \right| +$$

$$|f(b) \cdot \theta + f(b) \cdot \mu - f(x) \cdot [(1-\alpha)+\alpha]| \leq \frac{\epsilon}{4} + \frac{\epsilon}{4} + \frac{2\epsilon}{8} + \frac{2\epsilon}{8} = \epsilon$$

Therefore, for all $n \geq max\{n_1, n_2, n_3\}$ we obtain

$$|C_n(x) - f(x)| < \epsilon, \quad x \in [a, c]$$

2.3.2 Influence of the change of functions

Let f and g be continuous functions in $[a, b]$. The sequences $(C_{f,n})_n$ and $(C_{g,n})_n$ defined by (2.1) from f and g respectively, differ only at the values of the functions f and g at the points of the partitions P_n, that is to say,

$$C_{f,n}(x) = f(a) + \sum_{j=2}^{n} [f(x_j) - f(x_{j-1})] \cdot \frac{\sqrt[2n+1]{x_{j-1} - a} + \sqrt[2n+1]{x - x_{j-1}}}{\sqrt[2n+1]{b - x_{j-1}} + \sqrt[2n+1]{x_{j-1} - a}} \qquad (2.2)$$

$$C_{g,n}(x) = g(a) + \sum_{j=2}^{n} [g(x_j) - g(x_{j-1})] \cdot \frac{\sqrt[2n+1]{x_{j-1} - a} + \sqrt[2n+1]{x - x_{j-1}}}{\sqrt[2n+1]{b - x_{j-1}} + \sqrt[2n+1]{x_{j-1} - a}} \qquad (2.3)$$

2.3.3 Approximation in different practical cases

In this subsection we will use this method to approach a continuous function f from average samples of the values $f(x_j)$, from linear combinations of $f(x_j)$ and $f(x_{j+1})$ and from local average samples given by $\left(\chi_{[-\frac{h}{2}, \frac{h}{2}]} \star f \right)(x)$. Moreover, if the data $f(x_j)$ or average samples or local average samples are unknown, but approximate values of them are known, then we will prove that it is also possible to obtain the uniform reconstruction of the function f. Such problems often occur in environmental science, mathematical statistics, digital image, mechanics, numerical analysis and electricity; we refer to [Beh, Be2, Che, Del, Eps, Gon, Hua, Kil] for more details.

In this subsection, with the only condition that $f \in C[a, b]$, our purpose will be to employ the RAFU method to demonstrate that it is possible its reconstruction in all the mentioned cases. Moreover, the computational methods involved will be very easy to implement. So, we think that the importance of this technique can be useful to solve approximation problems in these practical situations.

2.3.3.1 Approximation from average samples

Proposition 2.2 *If the data $f(x_i)$, $i = 0, 1, ..., m$ which serve to define the step functions E_m as in Lemma (2.1) are substituted by $k_i = \frac{f(x_{i1})n_1 + ... + f(x_{ip})n_p}{n_1 + ... + n_p}$, $x_{1q} \in [a, x_1]$ or $x_{iq} \in (x_{i-1}, x_i]$, $i = 2, ..., n$, $q = 1, ..., p$, $n_1 + ... + n_q \neq 0$ then Theorem (2.1) holds.*

Proof 2.4 *From Propositions (1.3) and (1.4)*

If $n_i = 1$, we have the usual average values.

2.3.3.2 Approximation from local average samples

In many applications it is more realistic to assume that the available samples are local average samples near a certain x. We consider the special case in which we know data of the type

$$\left(\chi_{[-h,h]} \star f\right)(x) = \int_{-\infty}^{+\infty} \chi_{[-h,h]}(y) f(x - y) dy = \int_{x-h}^{x+h} f(z) dz \qquad (2.4)$$

where $\star$ denotes the convolution of the functions $\chi_{[-h,h]}$ and f. Sometimes we deal with phenomena which involve a function and its integral. For example, in mechanics, the velocity $v(t)$ and the displacement $s(t)$, or the acceleration $a(t)$ and the velocity $v(t)$. The tasks are to approximate the function f from integral values as (2.4).

Proposition 2.3 *If the data $f(x_i)$, $i = 0, 1, ..., m$ which serve to define the step functions E_m as in Lemma (2.1) are defined by $k_i = \frac{\int_{\tilde{x}_i - h}^{\tilde{x}_i + h} f(z) dz}{2h}$, with $[\tilde{x}_1 - h, \tilde{x}_1 + h] \subseteq [a, x_1]$ or $[\tilde{x}_i - h, \tilde{x}_i + h] \subseteq (x_{i-1}, x_i]$, $i = 2, ..., n$, then Theorem (2.1) holds.*

Proof 2.5 *From Propositions (1.3) and (1.4)*

2.3.3.3 Approximation from linear combinations

Proposition 2.4 *If the data $f(x_i)$, $i = 0, 1, ..., m$ which serve to define the step functions E_m as in Lemma (2.1) are replaced by $k_i = \frac{f(\tilde{x}_i) - f(\tilde{x}_{i-1})}{\tilde{x}_i - \tilde{x}_{i-1}} \cdot (x_i' - \tilde{x}_{i-1}) + f(\tilde{x}_{i-1})$ with $x_1' \in [\tilde{x}_0, \tilde{x}_1] \subseteq [a, x_1]$ or $x_i' \in [\tilde{x}_{i-1}, \tilde{x}_i] \subseteq (x_{i-1}, x_i]$, $i = 2, ..., n$, then Theorem (2.1) holds.*

Proof 2.6 *From Propositions (1.3) and (1.4)*

2.3.3.4 Approximation from approximate values

If $\{f(x_i)\}_{i=1}^{n}$ are unknown but we know the approximate data $\{f(x_i) + \eta_i\}_{i=1}^{n}$ with $|\eta_i| < \eta$ then, what is the difference in uniform norm between f and the sequence $(C_n)_n$ obtained from the approximate values? Next, we will answer this question.

Proposition 2.5 *Let f be a continuous function on $[a, b]$. For each n let $P_n = \{x_i\}_{i=0}^{n}$ be a partition of $[a, b]$ where $x_i = a + i \cdot \frac{b-a}{n}$, $i = 0, ..., n$. Suppose the values $\{f(x_i) + \eta_i\}_{i=0}^{n}$ are known and consider $|\eta_i| < \eta$ for $\eta > 0$. We construct the sequence $(C_n)_n$ defined by (2.1) but from the values $\{f(x_i) + \eta_i\}_{i=0}^{n}$. Then, for any $\epsilon > 0$ there exists n_0 such that*

$$|C_n(x) - f(x)| < \eta + \epsilon, \quad x \in [a, b]$$

for all $n > n_0$.

Proof 2.7 *Let $\epsilon > 0$ be fixed. Since f is uniformly continuous on $[a, b]$,*

$$\exists \delta > 0 : \mid x - x' \mid < \delta \Rightarrow \mid f(x) - f(x') \mid < \frac{\epsilon}{8}$$

From now on, we will make partitions of $[a, b]$ into intervals of length $< \delta$.

For each n such that $\frac{b-a}{n} < \delta$, we consider the step function

$$E_n(x) = (f(a) + \eta_1) \cdot \chi_{[x_0, x_1]} + \sum_{i=2}^{n} (f(x_i) + \eta_i) \cdot \chi_{(x_{i-1}, x_i]}$$

and the corresponding function $C_n(x)$ defined from E_n by the formula (2.1).

Proposition (1.3) holds when $n = m$. Therefore, fixed $\beta > 0$, there exists $n_1 \in \mathbb{N}$ such that if $n \geq n_1$ then

$$|C_n(x) - f(x)| \leq |C_n(x) - E_n(x)| + |E_n(x) - f(x)| < \eta + \epsilon$$

being $x \in [x_0, x_1 - \beta] \cup [x_1 + \beta, x_2 - \beta] \cup ... \cup [x_{n-1} + \beta, x_n]$.

Proposition (1.4) holds when $n = m$. Therefore, for $\beta > 0$ there exists $n_2 \in \mathbb{N}$ such that if $n \geq n_2$, $\alpha \in (0, 1)$ and $x \in (x_i - \beta, x_i + \beta)$, $i = 1, ..., n - 1$

$$|C_n(x) - f(x)|$$

$$\leq |C_n(x) - [(f(x_i) + \eta_i) \cdot (1 - \alpha) + (f(x_{i+1}) + \eta_{i+1}) \cdot \alpha]|$$

$$+ |[(f(x_i) + \eta_i) \cdot (1 - \alpha) + (f(x_{i+1}) + \eta_{i+1}) \cdot \alpha] - f(x)|$$

$$\leq \frac{\epsilon}{2} + |[(f(x_i) + \eta_i) \cdot (1 - \alpha) + (f(x_{i+1}) + \eta_{i+1}) \cdot \alpha] - f(x)|$$

Taking into account the uniform continuity of f and putting $f(x) = f(x) \cdot ((1 - \theta) + \theta)$ where $\theta \in (0, 1)$ is arbitrary

$$\leq |[(f(x_i) + \eta_i) \cdot (1 - \alpha) + (f(x_{i+1}) + \eta_{i+1}) \cdot \alpha] - f(x)|$$

$$\leq \frac{\epsilon}{2} + |f(x_i) \cdot (1 - \alpha) - f(x) \cdot (1 - \theta)| + |f(x_{i+1}) \cdot \alpha - f(x) \cdot \theta|$$

$$+ |\eta_i \cdot (1 - \alpha) + \eta_{i+1} \cdot \alpha)| \leq \frac{\epsilon}{2} + \frac{2\epsilon}{8} + \frac{2\epsilon}{8} + max\{\eta_i, \eta_{i+1}\} \leq \epsilon + \eta$$

Finally, considering $n \geq max \{n_1, n_2\}$ we obtain

$$|C_n(x) - f(x)| \leq \epsilon + \eta, \quad x \in [a, b]$$

Remark

The uniform stability of the RAFU method improves the instability of the interpolation by means of polynomials and this is a special contribution of this approximation procedure.

Proposition 2.6 *Let f be a continuous function in $[a, b]$ and let k be a real number. Let $(C'_n)_n$ be the sequence defined by (2.1) that converges uniformly to $f + k$ on $[a, b]$. Then, for any $\epsilon > 0$ there exists $n_0 \in \mathbb{N}$ such that if $n \geq n_0$*

$$|k| - \epsilon \leq \left| C'_n(x) - f(x) \right| \leq |k| + \epsilon, \quad x \in [a, b]$$

Proof 2.8 *For each $n \in \mathbb{N}$ let $P_n = \{a = x_0, ..., x_n = b\}$ be a partition. We can construct the sequences $(C'_n)_n$ and $(C_n)_n$ from the values $\{f(x_i) + \eta\}_{i=0}^{n}$ and $\{f(x_i)\}_{i=0}^{n}$ respectively as (2.1).*

It is clear that $C'_n(x) - C_n(x) = k$ on $[a, b]$ for any $n \in \mathbb{N}$.

For any $\epsilon > 0$ there exists $n_0 \in \mathbb{N}$ such that if $n \geq n_0$, $x \in [a, b]$,

$$\left| C'_n(x) - f(x) \right| \leq \left| C'_n(x) - C_n(x) \right| + |C_n(x) - f(x)| \leq |k| + \epsilon$$

by the uniform convergence of $(C_n)_n$ to f.

On the other hand

$$|k| = \left| C'_n(x) - C_n(x) \right| \leq \left| C'_n(x) - f(x) \right| + |f(x) - C_n(x)|$$

so that,

$$|k| - |f(x) - C_n(x)| \leq \left| C'_n(x) - f(x) \right|$$

Finally, for $n \geq n_0$ it results that

$$|k| - \epsilon \leq \left|C'_n(x) - f(x)\right| \leq |k| + \epsilon, \quad x \in [a, b]$$

2.4 Uniform approximation algorithms

In this section we consider continuous functions defined in $[a, b]$ and partitions $P_n = \{a = x_0, x_1, ..., x_{n-1}, x_n = b\}$ where $x_i = x_0 + \frac{b-a}{n} \cdot i$, $i = 0, ..., n$. We will use Mathematica ([Cas, Ram]) to construct four uniform approximation algorithms.

Algorithm 1. Uniform approximation to an arbitrary continuous function.

Suppose that the values $f(x_i)$, $i = 0, ..., n - 1$ are known.

Example 2.1 *Approximation to $f(x) = \frac{15 \cdot \sin x}{x^2 + 1} + \sqrt{x^2 + 2}$ in $[-5, 6]$ with C_{75}.*

In Figure (2.1) a) we represent the result.

$$f[x_-] := \frac{15 Sin[x]}{x^2 + 1} + \sqrt{x^2 + 2}; a = -5; b = 6; n = 75; h = \frac{b - a}{n};$$

$$t = Table\{a + jh, \{j, 0, n\}\}; k = Table\{f[a + jh], \{j, 0, n - 1\}\};$$

$$tt = Length[t]; kk = Length[k];$$

$$For[i = 2, i \leq kk, i++, M_i = \frac{(k_i - k_{i-1}) \sqrt[2n+1]{t_i - t_1}}{\sqrt[2n+1]{t_{tt} - t_i} + \sqrt[2n+1]{t_i - t_1}}];$$

$$For[i = 2, i \leq kk, i++, N_i = \frac{(k_i - k_{i-1})}{\sqrt[2n+1]{t_{tt} - t_i} + \sqrt[2n+1]{t_i - t_1}}];$$

$$f1[x_-] = k_1 + \sum_{i=2}^{kk} \left(M_i + N_i \sqrt[2n+1]{|x - t_i|} \cdot Sign\,(x - t_i) \right)$$

$$Plot[\{f[x], f1[x_-], \{x, t_1, t_{tt}\}]$$

Algorithm 2. Uniform approximation from average values.

Suppose that the data $\frac{f(x_{i,1}) + ... + f(x_{i,k})}{k}$ are known and $x_{i,j} \in [x_i, x_{i+1}]$ with $i = 0, ..., n - 1$ and $j = 1, ..., k$.

Example 2.2 *Approximation to* $f(x) = 10 + 25 \sin\left(\frac{x}{120}\right) + 3 \sin\left(\frac{x}{75}\right) + 10 \sin\left(\frac{x}{25}\right)$
on $[0, 165]$ *from averages of* f *at* 30 *points of each subinterval of length* $\frac{165-0}{900}$.
In Figure (2.1)) b) we represent the result.

$$f[x_-] := 10 + 25 \, Sin\left[\frac{x}{120}\right] + 3 \, Sin\left[\frac{x}{75}\right] + 10 \, Sin\left[\frac{x}{25}\right];$$

$$a = 0; b = 165; n = 900; h = \frac{b-a}{n}; v = \frac{n}{30};$$

$$t = Table\{a + 30hi, \{i, 0, v\}\}; d = Table\{f[a + jh], \{j, 0, n - 1\}\};$$

$$k = Table[\frac{\sum_{m=1}^{30} d_{m+30(i-1)}}{30}, \{i, 1, v\}];$$

$$tt = Length[t]; kk = Length[k];$$

$$For[i = 2, i \leq kk, i++, M_i = \frac{(k_i - k_{i-1}) \sqrt[2n+1]{t_i - t_1}}{\sqrt[2n+1]{t_{tt} - t_i} + \sqrt[2n+1]{t_i - t_1}}];$$

$$For[i = 2, i \leq kk, i++, N_i = \frac{(k_i - k_{i-1})}{\sqrt[2n+1]{t_{tt} - t_i} + \sqrt[2n+1]{t_i - t_1}}];$$

$$f1[x_-] = k_1 + \sum_{i=2}^{kk}\left(M_i + N_i \sqrt[2n+1]{|x - t_i|} \cdot Sign\left(x - t_i\right)\right)$$

$$Plot[\{f[x], f1[x]\}, \{x, t_1, t_{tt}\}]$$

Algorithm 3. Uniform approximation from approximate values.
Suppose that $\{f(x_i) + \eta_i\}_{i=0}^{n}$ are konwn and consider $|\eta_i| < \eta$ for some $\eta > 0$.

Example 2.3 *Approximation to* $g(x) = e^{-x^2}$ *on* $[-3, 3]$ *from the approximate values
given by* $f(x) = e^{-x^2} + \frac{sen x}{10}$ *(in this case* $|\eta_i| \leq \frac{1}{10} = \eta$*).*
In Figure (2.1) c) we represent the results.

24

$$f[x_-] := Exp\left[-x^2\right] + \frac{Sin[x]}{10}; g[x_-] = Exp\left[-x^2\right];$$

$$a = -3; b = 3; n = 200; h = \frac{b-a}{n};$$

$$t = Table\{a + jh, \{j, 0, n\}\}; k = Table\{f[a + jh], \{j, 0, n-1\}\};$$

$$tt = Length[t]; kk = Length[k]; p = 2n + 1;$$

$$For[i = 2, i \leq kk, i + +, M_i = \frac{(k_i - k_{i-1})\ ^{2n+1}\sqrt{t_i - t_1}}{^{2n+1}\sqrt{t_{tt} - t_i} + \ ^{2n+1}\sqrt{t_i - t_1}}];$$

$$For[i = 2, i \leq kk, i + +, N_i = \frac{(k_i - k_{i-1})}{^{2n+1}\sqrt{t_{tt} - t_i} + \ ^{2n+1}\sqrt{t_i - t_1}}];$$

$$d[x_-] = k_1 + \sum_{i=2}^{kk} \left(M_i + N_i\ ^{2n+1}\sqrt{|x - t_i|} \cdot Sign\,(x - t_i) \right)$$

$$Plot[|g[x] - d[x]|, \{x, t_1, t_{tt}\}]$$

Algorithm 4. Approximation by parts.

Approximation to $f \in C[a, c]$ from two approximations to f on $[a, b]$ and $[b, c]$ respectively.

Example 2.4 *Let $f(x) = sinx + 3$ be defined in $[-5, 5]$. With an approximation $C_{n,1}$ to f on $[-5, 0]$ and another $C_{n,2}$ on $[0, 5]$, we construct an approximation C_n to f on $[-5, 5]$.*

In Figure (2.2) we represent the result.

$$f[x_-] := Sin[x] + 3;$$

$$a = -5; b = 0; c = 5; n = 40; h1 = \frac{b-a}{n}; h2 = \frac{c-b}{n};$$

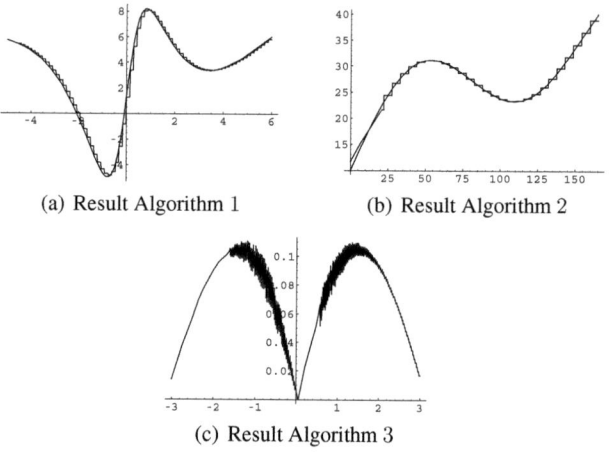

(a) Result Algorithm 1   (b) Result Algorithm 2

(c) Result Algorithm 3

Figure 2.1:

$$t1 = Table\{a + jh1, \{j, 0, n\}\}; k1 = Table\{f[a + jh1], \{j, 0, n - 1\}\};$$

$$tt1 = Length[t1]; kk1 = Length[k1];$$

$$For[i = 2, i \le kk1, i++, M1_i = \frac{(k1_i - k1_{i-1})}{\sqrt[2n+1]{c - t1_i} + \sqrt[2n+1]{t1_i - t1_1}} \sqrt[2n+1]{t1_i - t1_1}];$$

$$For[i = 2, i \le kk1, i++, N1_i = \frac{(k1_i - k1_{i-1})}{\sqrt[2n+1]{c - t1_i} + \sqrt[2n+1]{t1_i - t1_1}}];$$

$$For[i = 2, i \le kk1, i++, f1_i[x_-] := M1_i + N1_i \sqrt[2n+1]{|x - t1_i|} \cdot Sign[x - t1_i]];$$

$$f1[x_-] := k1_1 +$$

$$\sum_{i=2}^{kk1} f1_i[x] + \frac{(0 - k1_{kk1}) \left(\sqrt[2n+1]{t1_{tt1} - t1_1} + \sqrt[2n+1]{|x - t1_{tt1}|} \cdot Sign[x - t1_{tt1}]\right)}{\sqrt[2n+1]{c - t1_{tt1}} + \sqrt[2n+1]{t1_{tt1} - t1_1}};$$

$$Plot[\{f[x], f1[x]\}, \{x, t1_1, c\},$$

$$PlotStyle \rightarrow \{\{Thickness[0.005]\}, \{Thickness[0.01]\}\}]$$

$$t2 = Table\{a + jh2, \{j, 0, n\}\}; k2 = Table\{f[a + jh2], \{j, 0, n - 1\}\};$$

$$tt2 = Length[t2]; kk2 = Length[k2];$$

$$For[i = 2, i \leq kk2, i++, M2_i = \frac{(k2_i - k2_{i-1})\ \sqrt[2n+1]{t2_i - a}}{\sqrt[2n+1]{t2_{tt2} - t2_i} + \sqrt[2n+1]{t2_i - a}}];$$

$$For[i = 2, i \leq kk2, i++, N2_i = \frac{(k2_i - k2_{i-1})}{\sqrt[2n+1]{t2_{tt2} - t2_i} + \sqrt[2n+1]{t2_i - a}}];$$

$$For[i = 2, i \leq kk2, i++, f2_i[x_-] := M2_i + N2_i\ \sqrt[2n+1]{|x - t2_i|} \cdot Sign[x - t2_i]];$$

$$f2[x_-] := \frac{(k2_1 - 0)\left(\sqrt[2n+1]{t2_1 - a} + \sqrt[2n+1]{|x - t2_1|} \cdot Sign[x - t2_1]\right)}{\sqrt[2n+1]{t2_{tt2} - t2_i} + \sqrt[2n+1]{t2_i - a}} + \sum_{i=2}^{kk2} f2_i[x]$$

$$Plot[\{f[x], f2[x]\}, \{x, t1_1, c\},$$

$$PlotStyle \rightarrow \{\{Thickness[0.005]\}, \{Thickness[0.01]\}\}]$$

$$Plot[\{f[x], f1[x] + f2[x]\}, \{x, t1_1, c\},$$

$$PlotStyle \rightarrow \{\{Thickness[0.005]\}, \{Thickness[0.01]\}\},$$

$$PlotRange \rightarrow \{\{a, c\}, \{0, 4\}\}]$$

As an interesting exercise, the reader can change n, the interval or the function f and observe the results that Algorithms 1, 2, 3 and 4 show.

Moreover, depending on the problem, it is clear that one can construct other approximation algorithms from these ones and the results mentioned in the previous sections.

Example 2.5 *Approximation to* $f(x) = \frac{Sin^2(2x)+6}{20}$ *on* $[0, 8]$ *from linear combinations given by* $k_i = \frac{f(x_i)-f(x_{i-1})}{x_i-x_{i-1}} \cdot (x_i' - x_{i-1}) + f(x_{i-1})$ *with* $x_i' = \frac{x_i-x_{i-1}}{2}$, $i = 1, ..., n$,
In Figure (2.3) we represent the result.

Example 2.6 *Approximation to* $f(x) = \frac{Sin(20x)}{20}$ *on* $[0, 5]$ *from local average samples given by the* k_i *defined by the formula* $k_i = \frac{\int_{x_{i-1}}^{x_i} f(z)dz}{2}$, $i = 1, ..., n$,

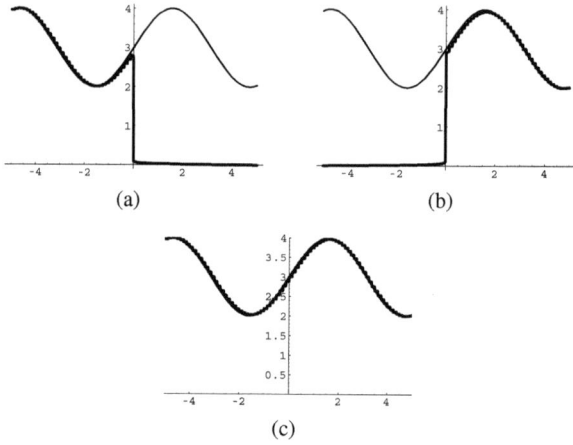

Figure 2.2:

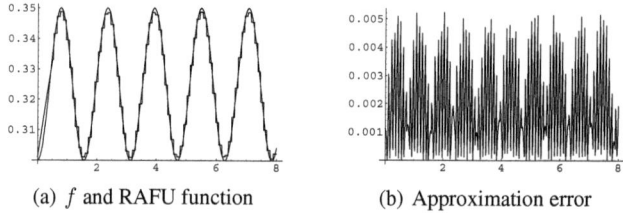

(a) f and RAFU function   (b) Approximation error

Figure 2.3:

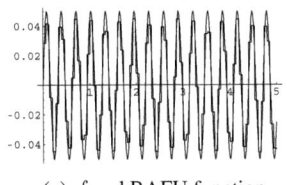
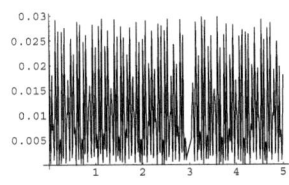

(a) f and RAFU function   (b) Approximation error

Figure 2.4:

In Figure (2.4) we represent the result.

2.5 The degree of approximation by RAFU functions

Definition 2.2 *Let f be a function defined in $[a, b]$. The modulus of continuity of f, $\omega(\delta)$, is the maximum of $|f(x) - f(y)|$ for all $a \leq x, y \leq b$, $|x - y| \leq \delta$.*

If f is continuous, $\omega(\delta)$ clearly decreases to 0 with δ. The degree of approximation of a function by arbitrary polynomials is given by the theorems of D. Jackson. (See, for example, [Lor], p.19-20.)

Theorem 2.2 *If $\omega(\delta)$ is the modulus of continuity of f in $[0, 1]$ then for each n there is a polynomial $P_n(x)$ of degree $\leq n$ such that*

$$|f(x) - P_n(x)| \leq C \cdot \omega(n^{-1})$$

where C is absolute constant. We may take for instance $C = 3$.

For the Bernstein polynomials, Sikkema ([Si1], [Si2]) proved that:

Theorem 2.3 *If $\omega(\delta)$ is the modulus of continuty of $f \in C[0, 1]$ then for each n it follows that*
$$|f(x) - B_n(f; x)| \leq 1,089887 \cdot \omega(n^{-\frac{1}{2}})$$

29

where $B_n(f;x)$ is the Bernstein polynomial of order n of f.

Esseen [Ess] also showed that

$$\lim_{n\to\infty} \sup_{0\leq x\leq 1} \max \frac{|f(x) - B_n(f;x)|}{w(n^{-\frac{1}{2}})} \leq 1.045564$$

For the RAFU method an analogous result can be proved. Next, we will show it.

Lemma 2.2 *It follows that:*

1. Let $1 \leq p \leq n-1$ be, $p \in \mathbb{N}$. Then $\left| {}^{2n-1}\!\!\sqrt{\frac{p}{n}} - 1 \right| \leq \frac{1}{3\sqrt{n}}$ *for $n \geq n_0$*

2. $\left| {}^{2n-1}\!\!\sqrt{\frac{1}{3}} - 1 \right| \leq \frac{1}{3\sqrt{n}}$ *for $n \geq n_0$*

3. $\left| {}^{2n-1}\!\!\sqrt{\frac{1}{3n}} - 1 \right| \leq \frac{1}{2\sqrt{n}}$ *for $n \geq n_0$*

4. Let $1 \leq p \leq n-1$ be, $p \in \mathbb{N}$. Then $\left| {}^{2n-1}\!\!\sqrt{n-p} - 1 \right| \leq \frac{\sqrt[7]{3}}{2\sqrt{n}}$ *for $n \geq n_0$*

It possible to check that in all parts of the previous lemma we can take $n_0 = 2$.

In this section we consider partitions $P_n = \{x_0, x_1, ..., x_n\}$ of $[a,b]$ where $x_i = x_0 + \frac{b-a}{n} \cdot i$, $i = 0,...,n$. Moreover, each interval $[x_{k-1}, x_k]$ of length $\frac{b-a}{n}$ will be divided into three equal parts of length $\frac{b-a}{3n}$:

$$\left[x_{k-1}, x_{k-1} + \tfrac{b-a}{3n}\right], \left[x_{k-1} + \tfrac{b-a}{3n}, x_k - \tfrac{b-a}{3n}\right], \left[x_k - \tfrac{b-a}{3n}, x_k\right]$$

Lemma 2.3 *Let P_n be a partition of $[a,b]$. For each $n \in \mathbb{N}$ and $k = 1, ..., n-1$ we define in $[a,b]$ the function*

$$F_n(x_k, x) = \frac{{}^{2n+1}\!\!\sqrt{x_k - a} + {}^{2n+1}\!\!\sqrt{x - x_k}}{{}^{2n+1}\!\!\sqrt{b - x_k} + {}^{2n+1}\!\!\sqrt{x_k - a}}$$

Then, if $x \in [a,b]$ and $k = 1, ...,n-1$ it verifies that $0 \leq F_n(x_k, x) \leq 1$.

The value of the function $F_n(x_k, x)$, for any k, does not depend on a or b. In fact, considering $x = a + \alpha_x \frac{b-a}{n}$ for an α_x, it follows that

$$F_n(x_k, x) = \frac{\sqrt[2n+1]{\left(a + k\frac{b-a}{n}\right) - a} + \sqrt[2n+1]{\left(a + \alpha_x\frac{b-a}{n}\right) - \left(a + k\frac{b-a}{n}\right)}}{\sqrt[2n+1]{\left(a + n\frac{b-a}{n}\right) - \left(a + k\frac{b-a}{n}\right)} + \sqrt[2n+1]{\left(a + k\frac{b-a}{n}\right) - a}} =$$

$$= \frac{\sqrt[2n+1]{k\frac{b-a}{n}} + \sqrt[2n+1]{(\alpha_x - k)\frac{b-a}{n}}}{\sqrt[2n+1]{(n-k)\frac{b-a}{n}} + \sqrt[2n+1]{k\frac{b-a}{n}}} = \frac{\sqrt[2n+1]{k} + \sqrt[2n+1]{(\alpha_x - k)}}{\sqrt[2n+1]{(n-k)} + \sqrt[2n+1]{k}}$$

Lemma 2.4 *Let P_n be a partition of $[a, b]$. Then for any $k = 1, ..., n-1$ and $x \in \left[x_{k-1} + \frac{b-a}{3n}, x_k - \frac{b-a}{3n}\right]$ it follows that*

1. *If $x - x_k > 0$ then $\dfrac{\sqrt[2n+1]{\frac{1}{n}} + \sqrt[2n+1]{\frac{1}{3n}}}{2} \leq F_n(x_k, x) \leq 1$*

2. *If $x - x_k < 0$ then $0 \leq F_n(x_k, x) \leq \dfrac{\sqrt[2n+1]{n-1} - \sqrt[2n+1]{\frac{1}{3}}}{2}$*

Moreover, these bounds are valid as $x \in \left[a, x_1 - \frac{b-a}{3n}\right]$, $x \in \left[x_{n-1} + \frac{b-a}{3n}, b\right]$ and $x \in \left(x_j - \frac{b-a}{3n}, x_j + \frac{b-a}{3n}\right)$ with $j \neq k$.

Lemma 2.5 *Let P_n be a partition of $[a, b]$. If $x \in \left[x_{k-1} + \frac{b-a}{3n}, x_k - \frac{b-a}{3n}\right]$, with $k = 1, ..., n-1$, $x \in \left[a, x_1 - \frac{b-a}{3n}\right]$, $x \in \left[x_{n-1} + \frac{b-a}{3n}, b\right]$ or $x \in \left(x_j - \frac{b-a}{3n}, x_j + \frac{b-a}{3n}\right)$ where $j \neq k$ then for all $n \geq 2$ it verifies that*

1. $\left| \dfrac{\sqrt[2n+1]{\frac{1}{n}} + \sqrt[2n+1]{\frac{1}{3n}}}{2} - 1 \right| \leq \dfrac{5}{12\sqrt{n}}$

2. $\left| \dfrac{\sqrt[2n+1]{n-1} - \sqrt[2n+1]{\frac{1}{3}}}{2} - 0 \right| \leq \dfrac{3\sqrt[7]{3} + 2}{12\sqrt{n}}$

Proofs of Lemmas (2.2), (2.3), (2.4) and (2.5) are obtained by elementary estimates.

Theorem 2.4 *Let P_n be a partition of $[a, b]$ and E_n the step function defined by*

$$E_n(x) = \begin{cases} k_1 & x \in [a, x_1] \\ k_2 & x \in (x_1, x_2] \\ ... & \\ k_n & x \in (x_{n-1}, b] \end{cases} \qquad k_j \in \mathbb{R}, \ j = 1, ..., n \qquad (2.5)$$

Then, for all $n \geq 2$ it follows that:

1. $|C_n(x) - E_n(x)| \leq \frac{M_n - m_n}{\sqrt{n}}$ *if* $x \in [a, b] \setminus \cup_{k=1}^{n-1} \left(x_k - \frac{b-a}{3n}, x_k + \frac{b-a}{3n} \right)$

2. $|C_n(x) - [k_j(1 - \alpha_x) + k_{j+1}\alpha_x]| \leq \frac{M_n - m_n}{\sqrt{n}}$ *when* $j = 1,..., n-1$ *and* $x \in \left(x_j - \frac{b-a}{3n}, x_j + \frac{b-a}{3n} \right)$

being M_n and m_n the maximum and the minimum of the k_j, $\alpha_x \in (0, 1)$ a number which depends upon x and C_n the radical function associated to E_n for each n.

The reader can see a complete proof in [Co1].

Corollary 2.2 *Let f be a continuous function defined in $[a, b]$. Then there exists a sequence of radical functions $(C_n)_n$ defined in $[a, b]$ such that*

$$|C_n(x) - f(x)| \leq \frac{M - m}{\sqrt{n}} + \omega \left(\frac{b - a}{n} \right)$$

for all $n \geq 2$, being M and m the maximum and the minimum of f in $[a, b]$ respectively and $\omega \left(\frac{b-a}{n} \right)$ its modulus of continuity.

Proof 2.9 *For each $n \geq 2$, let P_n be a partition of $[a, b]$, let E_n be the step function defined by*

$$E_n(x) = \begin{cases} f(a) & x \in [a, x_1] \\ f(x_2) & x \in (x_1, x_2] \\ ... \\ f(b) & x \in (x_{n-1}, b] \end{cases}$$

and let C_n be the corresponding radical function associated to E_n.
If $x \in [a, b] \setminus \cup_{k=1}^{n-1} \left(x_k - \frac{b-a}{3n}, x_k + \frac{b-a}{3n} \right)$ then,

$$|C_n(x) - f(x)| = |C_n(x) - E_n(x) + E_n(x) - f(x)|$$

$$\leq \frac{M_n - m_n}{\sqrt{n}} + |E_n(x) - f(x)| = \frac{M_n - m_n}{\sqrt{n}} + |f(x_j) - f(x)|$$

$$\leq \frac{M-m}{\sqrt{n}} + \omega\left(\frac{b-a}{n}\right)$$

Here, we take into account that $E_n(x) = f(x_j)$ for some j and Theorem (2.4).

If $x \in \cup_{k=1}^{n-1}\left(x_k - \frac{b-a}{3n}, x_k + \frac{b-a}{3n}\right)$ then, we can apply Theorem (2.4) and we can choose an appropiate index j to obtain

$$|C_n(x) - f(x)| \leq |C_n(x) - [f(x_j)(1-\alpha_x) + f(x_{j+1})\alpha_x]|$$

$$+ |[f(x_j)(1-\alpha_x) + f(x_{j+1})\alpha_x] - f(x)|$$

$$\leq \frac{M_n - m_n}{\sqrt{n}} + |[f(x_j)(1-\alpha_x) + f(x_{j+1})\alpha_x] - f(x)(1 \pm \alpha_x)|$$

$$\leq \frac{M-m}{\sqrt{n}} + |f(x_j) - f(x)|(1-\alpha_x) + |f(x_{j+1}) - f(x)|(1-\alpha_x)$$

$$\leq \frac{M-m}{\sqrt{n}} + \omega\left(\frac{b-a}{n}\right)(1 - \alpha_x + \alpha_x) = \frac{M-m}{\sqrt{n}} + \omega\left(\frac{b-a}{n}\right)$$

Example 2.7 *According to Corollary (2.2), here we approximate the function $f(x) = \sin(x)$ in the interval $[-3, 3]$ with the RAFU function C_{100}. The theoretical error given by Corollary (2.2) is $\frac{2}{\sqrt{100}} + \frac{6}{100} = 0.26$. In Figure (2.5) one can see the result.*

2.6 Comparative analysis of the degree of approximation

We will make an introductory analysis to compare the approximation by means of radical functions with the approximation by arbitrary polynomials and Bernstein polynomials.

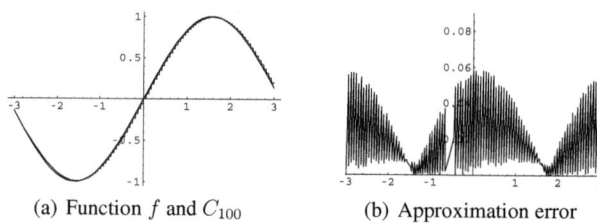

(a) Function f and C_{100} (b) Approximation error

Figure 2.5:

Suppose $f \in C\,[0,1]$. According to Theorem (2.2) and Corollary (2.2) the error with the RAFU functions is less than the error with arbitrary polynomials if the following inequality holds

$$2\omega\left(\frac{1}{n}\right) \geq \frac{M-m}{\sqrt{n}}$$

Example 2.8 *The function* $f(x) = \begin{cases} x \cdot sin\left(\frac{1}{x}\right) & x \neq 0 \\ 0 & x = 0 \end{cases}$ *is continuous in* $[0,1]$ *and it follows that*

$$2\sqrt{n}\omega\left(\frac{1}{n}\right) \geq 2\sqrt{n}\,|f(x) - f(y)| = 2\sqrt{n}\left|f'(x_0) \cdot \frac{1}{n}\right| =$$

$$2\sqrt{n}\left|sen\left(\frac{1}{x_0}\right) - \frac{1}{x_0} \cdot cos\left(\frac{1}{x_0}\right)\right| \cdot \frac{1}{n} =$$

$$2\sqrt{n}\,|0 - 2\pi n| \cdot \frac{1}{n} = 4\pi\sqrt{n} \geq M - m$$

taking values $x_0 = \frac{1}{2\pi n}$ *and considering appropiate points* x, y.

Definition 2.3 *Let* f *be a function defined in* $[a,b]$. *The function* f *belongs to the Lipschitz class, Lipα, if* $|f(x) - f(y)| \leq K\,|x - y|^\alpha$ *being* $a \leq x,y \leq b$, $0 < \alpha \leq 1$ *and* K *a real positive number.*

In the case of the arbitrary polinomials it is known (see [Lor]) that if f is defined in

34

$[0, 1]$ and $f \in Lip\alpha$, then for each n there is a polynomial P_n of degree n such that

$$|C_n(x) - P_n(x)| = O\left(n^\alpha\right)$$

A similar result can be obtained for the RAFU functions.

Theorem 2.5 *Let* $f \in Lip\alpha$ *be a function defined in* $[0, 1]$ *and* $K > 0$ *its lipschitz constant. Then, there exists a sequence of radical functions* $(C_n)_n$ *defined in* $[0, 1]$ *such that*

$$|C_n(x) - f(x)| \leq K \left[\frac{1}{\sqrt{n}} + \frac{1}{n^\alpha}\right], \qquad n \geq 2$$

Proof 2.10 *From Corollary (2.2) and the previous definitions.*

Suppose $f \in C[0, 1]$. In accordance with the result given by Essen [Ess] and Corollary (2.2), the error with the RAFU functions is less than the error with Bernstein polynomials if the following inequality holds

$$\frac{M - m}{\sqrt{n}} + \omega\left(\frac{1}{n}\right) \leq 1,045564 \cdot \omega(n^{-\frac{1}{2}})$$

If the function f has derivative, this inequality takes the form

$$\frac{M - m}{\sqrt{n}} + |f'(x_0)|\frac{1}{n} \leq 1,045564 \cdot |f'(y_0)|\frac{1}{\sqrt{n}}, \quad x_0, y_0 \in [0, 1]$$

That is to say

$$M - m + |f'(x_0)|\frac{1}{\sqrt{n}} \leq 1,045564 \cdot |f'(y_0)|, \quad x_0, y_0 \in [0, 1]$$

Example 2.9 *Let* $f(x) = sin20x$ *be in* $[0, 1]$. *In Figure (2.6) we show the results for* $n = 153$.

On the other hand, if f does not have derivative it is easy to find examples.

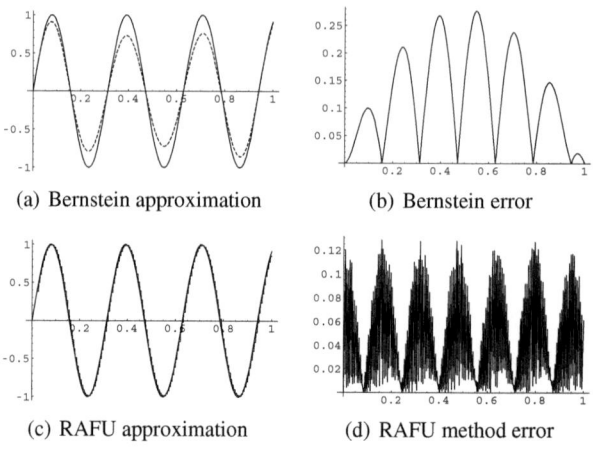

(a) Bernstein approximation   (b) Bernstein error

(c) RAFU approximation   (d) RAFU method error

Figure 2.6:

Example 2.10 *Let* $f(x) = |sin20x|$ *be in* $[0, 1]$. *In Figure (2.7) we show the results for* $n = 153$.

2.7 Uniform approximation from non uniform spaced points

In this subsection, we will consider partitions P_n of $[a, b]$ with non-uniformly spaced data.

Lemma 2.6 *Let* k *be a positive integer. Then, for* $n \geq 2$ *it verifies that*

1. $\left| \sqrt[2n+1]{n^k} - 1 \right| \leq \frac{(2k-1)\sqrt[7]{3}}{2\sqrt{n}}$

2. $\left| \sqrt[2n+1]{\frac{1}{n^k}} - 1 \right| \leq \frac{k}{3\sqrt{n}}$

Proof 2.11 *By induction on* k. *Cases* $k = 1$ *are in [Co1]. The proof finishes taking*

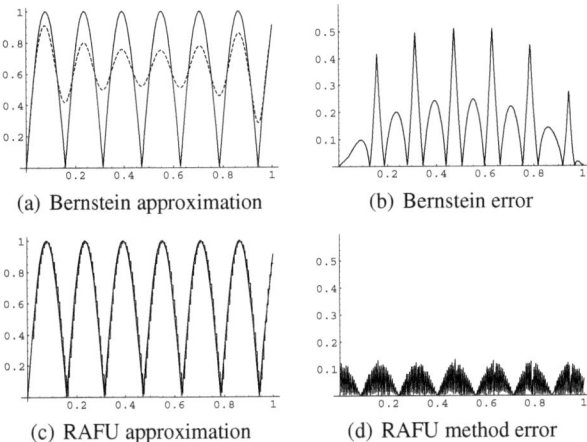

(a) Bernstein approximation   (b) Bernstein error

(c) RAFU approximation   (d) RAFU method error

Figure 2.7:

into account that

$$\left| \sqrt[2n+1]{n^{\pm k}} - 1 \right| = \left| \sqrt[2n+1]{n^{\pm k}} - \sqrt[2n+1]{n^{\pm 1}} + \sqrt[2n+1]{n^{\pm 1}} - 1 \right|$$

Lemma 2.7 *Let* $P_n = \{a = x_0, x_1, ..., x_s = b\}$ *be a partition of* $[a, b]$ *with* $\delta(s) = \min_{1 \leq j \leq s} |x_j - x_{j-1}|$. *Then, for any* $k = 1, ..., s-1$ *and* $x \in [a, b] \setminus \left(x_k - \frac{\delta(s)}{3}, x_k + \frac{\delta(s)}{3} \right)$ *it follows that:*

1. $\sqrt[2n+1]{\frac{\delta(s)}{b-a}} \frac{1 + \sqrt[2n+1]{\frac{1}{3}}}{2} \leq F_n(x_k, x) \leq 1$ *if* $x - x_k > 0$

2. $0 \leq F_n(x_k, x) \leq \frac{\sqrt[2n+1]{\frac{b-a}{\delta(s)}} - \sqrt[2n+1]{\frac{1}{3}}}{2}$ *if* $x - x_k < 0$

The proof can be obtained by elementary estimates.

Lemma 2.8 *Let* $K \geq 2$ *be a positive integer such that* $\frac{3(b-a)}{n^K} \leq \delta(s)$. *Then, for all* $n \geq 2$, *it verifies that*

1. $\left| 1 - \sqrt[2n+1]{\frac{\delta(s)}{b-a}} \frac{1 + \sqrt[2n+1]{\frac{1}{3}}}{2} \right| \leq \frac{K}{3\sqrt{n}}$

2. $\left| \dfrac{\sqrt[2n+1]{\frac{b-a}{\delta(s)}} - \sqrt[2n+1]{\frac{1}{3}}}{2} - 0 \right| \leq \dfrac{(6K-3)\sqrt[7]{3}+2}{12\sqrt{n}}$

Moreover, $max\left\{ \dfrac{K}{3}, \dfrac{(6K-3)\sqrt[7]{3}+2}{12} \right\} \leq \dfrac{3K}{5}$

The proof can be obtained easily from Lemma (2.6).

Proposition 2.7 *Let* $P_s = \{a = x_0, x_1, ..., x_s = b\}$ *be a partition of* $[a,b]$ *and let* E_s *be a step function defined in* $[a,b]$ *by*

$$E_s(x) = k_1 \cdot \chi_{[x_0,\, x_1]} + \sum_{i=2}^{s} k_i \cdot \chi_{(x_{i-1},\, x_i]} \quad k_i \in \mathbb{R}$$

If $\frac{3(b-a)}{n^K} \leq \delta(s)$, *being* $\delta(s) = \min\limits_{1\leq j\leq s} |x_j - x_{j-1}|$ *and* $K \geq 2$ *a positive integer, then for all* $n \geq 2$ *it follows that:*

1. *For all* $x \in [a,b] \setminus \cup_{j=1}^{s-1}\left(x_j - \frac{\delta(s)}{3}, x_j + \frac{\delta(s)}{3} \right)$

$$|C_n(x) - E_s(x)| \leq \frac{6K}{5} \frac{M_s - m_s}{\sqrt{n}}$$

2. *For all* $x \in \left(x_j - \frac{\delta(s)}{3}, x_j + \frac{\delta(s)}{3} \right)$, $j = 1, ..., s-1$

$$|C_n(x) - [k_j(1 - \alpha_x) + k_{j+1}\alpha_x]| \leq \frac{6K}{5} \frac{M_s - m_s}{\sqrt{n}}$$

where M_s *and* m_s *are the maximum and the minimum of the* k_j, $\alpha_x \in (0,1)$ *is a number which depends only on* x *and* $(C_n)_n$ *is the sequence of RAFU functions associated to* E_s.

Proof 2.12 *It is analogous to the proof given in [Co1] (p. 115-117) but now we use Lemmas (2.6), (2.7) and (2.8).*

Theorem 2.6 *Let $P_n = \{a = x_0, x_1, ..., x_{s_n} = b\}$ be a partition of $[a, b]$ with $\delta(s_n) = \min\limits_{1 \leq j \leq s_n} |x_j - x_{j-1}|$ and $\Delta(s_n) = \max\limits_{1 \leq j \leq s_n} |x_j - x_{j-1}|$ such that $\frac{3(b-a)}{n^K} \leq \delta(s_n) \leq \Delta(s_n) \leq h$ being $h = \frac{b-a}{n}$ and $K \geq 2$ a positive integer. Let f be a continuous functions in $[a, b]$, then there exists a sequence $(C_n)_n$ defined in $[a, b]$ such that*

$$\|f - C_n\| \leq \frac{6K}{5} \frac{M - m}{\sqrt{n}} + \omega(\Delta(s_n))$$

being $n \geq 2$, M and m the maximum and the minimum of f in $[a, b]$ respectively and $\omega(\Delta(s_n))$ its modulus of continuity, C_n as usual and $F_n(x_p, x) = \frac{\sqrt[2n+1]{x_p - a} + \sqrt[2n+1]{x - x_p}}{\sqrt[2n+1]{b - x_p} + \sqrt[2n+1]{x_p - a}}$, $p = 1, ..., s_n - 1$.

Proof 2.13 *The proof is the same as Corollary (2.2) but here we consider Lemmas (2.6), (2.7), (2.8) and Proposition (2.7).*

Example 2.11 *Theorem (2.6) is used to approach the function $f(x) = sin(x)$ in $[0, 0.5]$ by considering $n = 10$, $K = 2$ and $s_n = 15$. In this example we have used the values of f at the points of the partition P_n defined by*

$$\{0, 0.02, 0.05, 0.09, 0.14, 0.17, 0.21, 0.26, 0.31, 0.34, 0.39, 0.42, 0.44, 0.48, 0.5\}$$

These points verify that $0.015 \leq \delta(s_n) \leq \Delta(s_n) \leq 0.05$. Here, the theoretical error given by Theorem (2.6) is 0.463858. In Figure (2.8) we can see the graphical result.

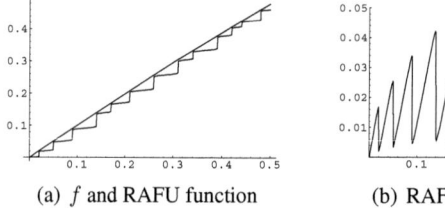

(a) f and RAFU function   (b) RAFU method error

Figure 2.8:

Chapter 3

The topological foundation of the RAFU method

3.1 Introduction

Let K be a compact Hausdorff space. The Kakutani-Stone Theorem [Kak] gives a necessary and sufficient condition for the density of a lattice of $C(K)$ in the topology of the uniform convergence on K. The Stone-Weierstrass Theorem [Sto] provides a necessary and sufficient condition under which an algebra of $C(K)$ is uniformly dense. Nevertheless, the above conditions are not sufficient to ensure the uniform density of a linear space of $C(K)$. Tietze [Tie], Jameson [Jam], Mrowka [Mro], Blasco-Moltó [Bla], Garrido-Montalvo [Mon] and Gassó-Hernández-Rojas [Gas] have studied the uniform approximation for linear spaces.

In Section 2 we will construct a RAFU linear space, $\mathcal{C}$, in $C[a, b]$ and we will prove that $\mathcal{C}$ is uniformly dense in $C[a, b]$ by using a S-separation condition due to Blasco-Moltó [Bla]. We will also see that the uniform density of $\mathcal{C}$ in $C[a, b]$ is not a consequence of the results given by Kakutani-Stone, Stone-Weierstrass, Tietze, Jameson, or Mrowka.

It is true that Blasco-Moltó showed an example of a linear space, $\mathcal{F}$, uniformly dense in $C\,[0,1]$ by using the S-separation condition but some questions were not studied: the linear combinations of elements belonging to $\mathcal{F}$ which approximate uniformly every $f \in C\,[0,1]$ and the degree of uniform approximation that $\mathcal{F}$ provides were unknown. In the Section 3 we will solve these problems by using the RAFU linear space C. Moreover, this linear space C can be used as an example of approximation by series in the work of Gassó-Hernández-Rojas.

The Section 3 is devoted to study the *best* approximation problem.

3.2 A RAFU linear space uniformly dense in $C\,[a,b]$

For each $n \in \mathbb{N}$, we consider the partition $P = \{x_0, x_1, ..., x_n\}$ of $[a,b]$ with $x_j = a + j \cdot \frac{b-a}{n}$, $j = 0, ..., n$ and we define in $[a,b]$ the functions

$$C_n(x) = k_1 + \sum_{i=2}^{n} (k_i - k_{i-1}) \cdot F_n(x_{i-1}, x) \tag{3.1}$$

where $\{k_i\}_{i=1}^{n}$ are a family of real arbitrary numbers and

$$F_n\,(x_{i-1}, x) = \frac{\sqrt[2n+1]{x_{i-1} - x_0} + \sqrt[2n+1]{x - x_{i-1}}}{\sqrt[2n+1]{x_n - x_{i-1}} + \sqrt[2n+1]{x_{i-1} - x_0}}, \quad i = 2, ..., n \tag{3.2}$$

In Section 2.2 we proved that the set C^* is uniformly dense in $C\,[a,b]$ but unfortunately this set is not a linear space. If C_p and C_q are two elements belonging to C^* then it is easy to check that $C_p + C_q \notin \mathsf{C}^*$.

Remark According to this, we made a mistake in [Co1] and [Co2]. The set C mentioned in these papers is not a linear space. But all the results mentioned in [Co1] and [Co2] are true if we change appropriately the definition of the set that contains the subsets C_n. Next, we do it.

As in Section 2.2, we designate by C_n the subset of $C\,[a,b]$ formed by the functions C_n but now we will denote by C the set $\mathsf{C} = \sum_{n\geq 2} \mathsf{C}_n$.

Proposition 3.1 *The sets C_n, $n \geq 2$, and C are linear spaces included in $C\,[a,b]$.*

Proof 3.1 *It is clear that $\mathfrak{C}_n$ and $\mathfrak{C}$ are included in $C[a,b]$. In the first place it is easy to check that $\mathfrak{C}_n$ is a n-dimensional linear space because n is fixed and hence the values $\{x_i\}_{i=0}^{n}$ are the same points. Moreover, a basis of $\mathfrak{C}_n$ is*

$$\{1, F_n(x_1,x), ..., F_n(x_{n-1},x)\}$$

$\mathfrak{C}$ *is a linear space by definition. A basis of $\mathfrak{C}$ is*

$$\{1, F_2(x_1,x), F_3(x_1,x), F_3(x_2,x), ..., F_n(x_1,x), ..., F_n(x_{n-1},x), ...\}$$

Definition 3.1 *A RAFU linear space is a linear space whose basis is formed by radical functions of the type (3.2). We will say that $\mathfrak{C}$ is a RAFU linear space.*

The theorems of uniform approximation in $C(K)$ for lattices are known as Kakutani-Stone's theorems (the interested reader can see [Kak], [Sto], [St2]).

The family $\mathfrak{C}$ is not a lattice. In fact, in the interval $[-1,1]$ the function $C(x) = \sqrt[3]{x} \in \mathfrak{C}$ but $|C(x)| \notin \mathfrak{C}$ because at $x = 0$ its side derivatives do not have the same sign. Therefore, the family $\mathfrak{C}$ does not satisfy the Kakutani-Stone's theorems.

The theorems of uniform approximation in $C(K)$ for algebras are known as Stone-Weierstrass' theorems (the interested reader can see [Sto], [St2]).

A simple count proves that $\mathfrak{C}$ is not an algebra, therefore the set $\mathfrak{C}$ does not verify the Stone-Weierstrass' theorems.

Let X be a topological space and let $C^*(X)$ be the set consisting of all bounded continuous functions and let $C(X)$ be the set consisting of all continuous functions.

Definition 3.2 *Let $\mathcal{F}$ be a family of $C^*(X)$. We say that*

1. A zero-set in X is a set of the form $Z(f) = \{x \in X : f(x) = 0\}$ with $f \in C^*(X)$.

2. The Lebesgue-sets of $f \in C(X)$ are the sets defined as

$$L_\alpha(f) = \{x \in X : f(x) \le \alpha\}$$

43

and

$$L^\beta(f) = \{x \in X : f(x) \geq \beta\}$$

where α and β are real numbers.

3. *$\mathcal{F}$ S_1- separates the subsets A and B of X when there is $f \in \mathcal{F}$, $0 \leq f \leq 1$ such that $f(x) = 0$ if $x \in A$ and $f(x) = 1$ if $x \in B$.*

4. *(Blasco-Moltó [Bla]). $\mathcal{F}$ S-separates the subsets A and B of X if for each $\delta > 0$, there is $f \in \mathcal{F}$ such that $0 \leq f \leq 1$ for every $x \in X$, $f(A) \subset [0, \delta]$ and $f(B) \subset [1 - \delta, 1]$.*

5. *(Garrido-Montalvo [Mon]). $\mathcal{F}$ S'-separates the subsets A and B of X if for each $\delta > 0$, there is $f \in \mathcal{F}$ such that $-\delta \leq f \leq 1 + \delta$ for every $x \in X$, $f(A) \subset [-\delta, \delta]$ and $f(B) \subset [1 - \delta, 1 + \delta]$.*

6. *Given a series of continuous functions $\sum_{i \in I} f_i$ on X, the series is locally convergent, for every $x \in X$, if there is a neighborhood U of x such that the series converges uniformly on U. For $E \subset C(X)$, $\sum(E)$ is the set of all $f \in C(X)$ such that $f = \sum_{i \in I} f_i$ with $f_i \in E$ for every $i \in I$ and $\sum_{i \in I} f_i$ is a locally convergent series. $\overline{\sum(E)}$ denotes the uniform closure of $\sum(E)$.*

Theorem 3.1 *(Tietze [Tie], Mrowka [Mro]). Let $\mathcal{F}$ be a linear space of $C^*(X)$. $\mathcal{F}$ is uniformly dense in $C^*(X)$ if and only if $\mathcal{F}$ S_1- separates every pair of disjoint zero-sets in X.*

Theorem 3.2 *(Jameson [Jam]). Let $\mathcal{F}$ be a linear space of $C^*(X)$. $\mathcal{F}$ is uniformly dense in $C^*(X)$ if and only if $\mathcal{F}$ S_1- separates every pair of disjoint closed subsets in X.*

By the properties of the functions of the linear space $\mathcal{C}$ it is possible to deduce that we cannot apply to $\mathcal{C}$ the results of Tietze, Mrowka or Jameson.

Theorem 3.3 *(Blasco-Moltó [Bla]). Let X be a topological space. A linear space $\mathcal{F}$*

of $C^*(X)$ is uniformly dense in $C^*(X)$ if and only if $\mathcal{F}$ S- separates every pair of disjoint zero-sets in X.

We can apply this theorem to prove the uniform density of C in $C[a, b]$.

Theorem 3.4 *The RAFU linear space C is uniformly dense in the set $C[a, b]$.*

The reader can see a complete proof in [Co2].

The S-separation of subsets is equivalent to the S'-separation of subsets in linear spaces containing constant functions (Garrido-Montalvo [Mon]). Clearly C contains the constant functions, therefore we can also deduce the uniform density of C in $C[a, b]$ by using the S'-separation condition of every pair of disjoint *zero-sets* in X.

Blasco-Moltó [Bla] proved that the linear subspace $\mathcal{F}$ of $C[0, 1]$ generated by the functions

$$\{exp((x + \mu)^n) : \mu \in \mathbb{R}, x \in [0, 1], n = 0, 1, 3, ..., 2k + 1, ...\}$$

is uniformly dense in $C[0, 1]$, but the linear combinations which approximate uniformly a function $f \in C[0, 1]$ and the degree of uniform approximation that $\mathcal{F}$ provides were not studied. In the case of the RAFU method we have obtained both the linear combinations that approximate a function $f \in C[0, 1]$ and the degree of uniform approximation that C provides and this is an interesing improvement.

Theorem 3.5 *(Gassó-Hernández-Rojas [Gas]). Let A be a subset of $C(X)$ and let E be a linear space of $C(X)$ which S-separates Lebesgue-sets of A. Then the sublattice generated by A is contained in $\overline{\sum(E)}$.*

The RAFU linear space C satifies Theorem (3.5) when $X = [a, b]$ because we have proved that C S-separates every pair of disjoint *zero-sets* Z_1 and Z_2 of $[a, b]$ and every *Lebesgue-set* is also a *zero-set* since $L_\alpha(f) = Z((f - \alpha) \vee 0)$ and $L^\beta(f) = Z((f - \beta) \wedge 0)$. In this case, if $A = C(X)$ we can say that $C(X)$ is contained in $\overline{\sum(\mathsf{C})}$. In fact, given $f \in C[a, b]$, we already knew that $f(x) = \sum_{n=1}^{\infty} c_n(x)$ where $c_1 = f(a)$, $c_n \in \mathsf{C}$, $n \geq 2$, and the series converges uniformly.

3.3   The *best* approximation problem

Let $\mathcal{F}$ be a n-linear subspace of $C\,[a,b]$ and $f \in C\,[a,b]$. A function $u \in \mathcal{F}$ which minimizes

$$\|f(x) - u(x)\|_\infty = \max_{a \leq x \leq b} |f(x) - u(x)|$$

is conventionally called a function of *best* approximation to f in $\mathcal{F}$.

Definition 3.3 *Let $\mathcal{F}$ be a n-linear subspace of $C\,[a,b]$. We say that $\mathcal{F}$ satisfies the Haar property if the only function of $\mathcal{F}$ with m different zeros is $f = 0$.*

It is well known the following result.

Theorem 3.6 *Let $\mathcal{F}$ be a n-linear subspace of $C\,[a,b]$ and $f \in C\,[a,b]$. Then there exists a unique uniform approximation to f on $\mathcal{F}$ if and only if $\mathcal{F}$ satisfies the Haar condition.*

Studying the monotony of each C_n, one can check that any $C_n \in \mathsf{C}_n$ has at the most $n - 1$ zeros in $C\,[a,b]$, being $P_n = \{x_i\}_{i=0}^n$ where $x_j = a + j \cdot \frac{b-a}{n}$, $j = 0, ..., n$. Hence, as polynomials and trigonometric sums, the RAFU linear space C also verifies the Haar property.

The next theorem characterizes the *best* approximation.

Theorem 3.7 *Let $\mathcal{F}$ be a n-linear subspace of $C\,[a,b]$ satisfying the Haar property, $f \in C\,[a,b]$ and $u \in \mathcal{F}$. Then u is the best uniform approximation to f on $\mathcal{F}$ if and only if there are $n + 1$ points $x_i \in [a,b]$, $i = 0, ..., n$, such that $x_0 < x_1 < ... < x_n$ and*

$$f(x_j) - u(x_j) = (-1)^j M, \quad j = 0, ..., n$$

with $M = \max_{x \in [a,b]} |f(x) - u(x)|$ or $-M = \max_{x \in [a,b]} |f(x) - u(x)|$.

For any subspace $\mathcal{F}$, this result shows the difficulty to find the *best* uniform approximation to a continuous function f and this same difficulty occurs as $\mathcal{F} = \mathsf{C}_n$.

However, we show the following uniform approximation procedure that is based on the arguments of the previous theorem.

For each $i = 1, ..., n$, we define

$$m_i = min\left\{f(x) : x \in [x_{i-1}, x_i]\right\}$$

$$M_i = max\left\{f(x) : x \in [x_{i-1}, x_i]\right\}$$

$$m = min\left\{m_i : i = 1, ..., n\right\} = min\left\{f(x) : x \in [a, b]\right\}$$

$$M = max\left\{M_i : i = 1, ..., n\right\} = max\left\{f(x) : x \in [a, b]\right\}$$

and the step function

$$h(x) = k_1 \cdot \chi_{[x_0, x_1]} + \sum_{i=2}^{n} k_i \cdot \chi_{(x_{i-1}, x_i]}$$

where each $k_i = \frac{m_i + M_i}{2}$ is the *best* uniform approximation to f on $[x_{i-1}, x_i]$ if we take the set of all constant functions. For this step function, we consider its associated sequence of radical functions $(C_n)_n$ defined by the formula

$$C_n(x) = k_1 + \sum_{i=2}^{n} (k_i - k_{i-1}) \cdot F_n(x_{i-1}, x)$$

In this way, according to Corollary (2.2) we obtain that

$$\|C_n - f\| \leq \|C_n - h\| + \|h - f\| \leq \frac{M - m}{\sqrt{n}} + \omega\left(\frac{b - a}{n}\right)$$

Chapter 4

Approximation in different smoothness spaces with the RAFU method

4.1 Introduction

On the one hand, we have proved in Chapter 1 that we can use the RAFU method to approach an arbitrary step function. From this result, we have demonstrated in Chapter 2 that the RAFU functions are dense in $C[a, b]$. On the other hand, it is well-known that the step functions are dense in other spaces of functions, so our aim in this chapter will be to prove that these RAFU functions can also be dense in all these spaces.

In Section 2 we will approach functions of the spaces $C_0(\mathbb{R})$ and $C_{00}(\mathbb{R})$. The Riemann integrable functions will be approximated with the RAFU method in Section 3. Moreover, the integral of a Riemann integrable function will be approximated by the sequence of integrals of the functions that the RAFU method provides. This approximation procedure will serve to approach the Lebesgue integrable functions in Section 4. Section 5 is devoted to approximate elements of $L^p[a, b]$ and $L^p(\mathbb{R})$, $1 \leq p < \infty$. In Section 6 the RAFU method will be employed to approach measurable functions.

4.2   Approximation on $C_0 (\mathbb{R})$ and $C_{00} (\mathbb{R})$

Definition 4.1 $C_0 (\mathbb{R})$ *is the space of all continuous functions on $\mathbb{R}$ such that $\lim\limits_{|x| \to \infty} f(x)$*

exists and equals 0.

$C_{00} (\mathbb{R})$*is the space that consists of those functions on $\mathbb{R}$ with compact support.*

Note that the functions C_n defined on $[a, b]$ as (1.1) can be defined on $\mathbb{R}$ by the same formula. In this case, it is easy to check that these functions verify $\lim\limits_{n \to \infty} C_n(x) = f(a)$ for all $x \in (-\infty, a]$ and $\lim\limits_{n \to \infty} C_n(x) = f(b)$ for all $x \in [b, +\infty)$.

If $f \in C_{00} (\mathbb{R})$ there exists $M > 0$ such that $f(x) = 0$ if $|x| \geq M$. Then the sequence $(C_n)_n$, defined on $\mathbb{R}$ as (1.1) being $a = -M$ and $b = M$, verifies that $\lim\limits_{n \to \infty} C_n = f$ uniformly on $[-M, M]$ and $\lim\limits_{n \to \infty} C_n(x) = f(x) = 0$ for all $|x| \geq M$.

Let f be an element of $C_0 (\mathbb{R})$. Given $\epsilon > 0$, there exists N_ϵ such that $|f(x)| < \epsilon$ if $|x| > N_\epsilon$. For these $\epsilon > 0$ and N_ϵ, there is a function $C_{n,\epsilon}$, defined on $\mathbb{R}$ as (1.1) by considering $a = -N_\epsilon$, $b = N_\epsilon$ and by requiring $C_{n,\epsilon}$ to have the values 0 at the points $\pm N_\epsilon$, such that $|f - C_{n,\epsilon}| < \epsilon$ on $[-N_\epsilon, N_\epsilon]$. Thus, we can construct a sequence $(C_n)_n$ defined on $\mathbb{R}$ such that $\lim\limits_{n \to \infty} C_n(x) = f(x)$ for all $x \in \mathbb{R}$. Moreover, this limit becomes uniform on each interval $[-N_\epsilon, N_\epsilon]$.

4.3   Approximation to a Riemman integrable function

Proposition 4.1 *Let f be a bounded real-valued function on $[a, b]$. If f is Riemann integrable on $[a, b]$, then there is a sequence of radical functions $(C_n)_n$, defined as (1.1), such that $\lim\limits_{n \to \infty} C_n = f$ uniformly on $[a, b]$ except in a null set D.*

Proof 4.1 *The proof takes into account that it is well-known that there is a sequence of step functions $(E_m)_m$ defined on $[a, b]$ which converges uniformly to f on $[a, b]$ except in a null set D that contains the points in which f is not continuous. For details, the interested reader can see [Co3].*

Now we suggest the following approximation to the integral of a Riemann integrable function f by using the RAFU method.

Lemma 4.1 *Let E_m be a step function on $[a,b]$ defined as (2.5). Then, the sequence $(C_{m,n})_n$ defined from E_m verifies that*

$$\lim_{n\to\infty} \int_a^b C_{m,n} = \int_a^b E_m$$

.

Proof 4.2

$$\lim_{n\to\infty} \int_a^b C_{m,n}(x)dx$$

$$= \lim_{n\to\infty} \int_a^b \left(k_1 + \sum_{p=2}^m [k_p - k_{p-1}] \cdot F_n(x_{p-1}, m, x) \right)(x)dx$$

$$= \lim_{n\to\infty} \left(k_1 \cdot (b-a) + \sum_{p=2}^m \left(\frac{(k_p - k_{p-1}) \cdot (b-a) \cdot \sqrt[2n+1]{x_{p-1} - a}}{\sqrt[2n+1]{b - x_{p-1}} + \sqrt[2n+1]{x_{p-1} - a}} \right) + \frac{2n+1}{2n+2} \right.$$

$$\left. \cdot \sum_{p=2}^m \left(\frac{(k_p - k_{p-1}) \cdot \left[\sqrt[2n+1]{(b - x_{p-1})^{2n+2}} - \sqrt[2n+1]{(a - x_{p-1})^{2n+2}} \right]}{\sqrt[2n+1]{b - x_{p-1}} + \sqrt[2n+1]{x_{p-1} - a}} \right) \right)$$

$$= k_1 \cdot (b-a) + \sum_{p=2}^m \left(\frac{(k_p - k_{p-1}) \cdot (b-a)}{2} \right)$$

$$+ \sum_{p=2}^m \left(\frac{(k_p - k_{p-1}) \cdot (b - 2x_{p-1} + a)}{2} \right)$$

$$= \sum_{i=1}^m k_i \cdot (x_i - x_{i-1}) = \int_a^b E_m$$

The expression

$$I_n(C_{m,n})$$

$$= \left[k_1 + \sum_{p=2}^{m} \left(\frac{(k_p - k_{p-1}) \cdot \sqrt[2n+1]{x_{p-1} - a}}{\sqrt[2n+1]{b - x_{p-1}} + \sqrt[2n+1]{x_{p-1} - a}} \right) \right] \cdot (b - a) + \frac{2n + 1}{2n + 2}$$

$$\cdot \left(\sum_{p=2}^{m} \frac{(k_p - k_{p-1}) \cdot \left[\sqrt[2n+1]{(b - x_{p-1})^{2n+2}} - \sqrt[2n+1]{(a - x_{p-1})^{2n+2}} \right]}{\sqrt[2n+1]{b - x_{p-1}} + \sqrt[2n+1]{x_{p-1} - a}} \right)$$

can be considered a new formula of numerical approximation to the integral $\int_a^b E_m$. Thus, if f is a Riemann integrable function, we can approach $\int_a^b f$ taking into account that

$$\int_a^b f = \lim_{m \to \infty} \lim_{n \to \infty} [I_n (C_{m,n})]$$

Example 4.1 *Let $f(x) = \frac{1}{x^2+1}$ be defined in $[0, 3]$. Its exact integral is*

$$\int_0^3 f(x)dx = 1.24905$$

and its approximate value for $n = 1990$ is

$$I_{1990} (C_{1990,1990}) = 1.24982$$

Example 4.2 *Lef f be the step function defined in $[0, 5]$ by*

$$f(x) = \begin{cases} 10 & if \quad 0 \le x \le 1 \\ -6 & if \quad 1 < x \le 2 \\ 4 & if \quad 2 < x \le 3 \\ -3 & if \quad 3 < x \le 4 \\ 2 & if \quad 4 < x \le 5 \end{cases}$$

Its exact integral is

$$\int_0^5 f(x)dx = 7$$

and its approximate value for $n = 1000$ *is*

$$I_{1000} \left(C_{1000,1000} \right) = 7.0115$$

4.4 Approximation to a Lebesgue integrable function

Let I be an arbitrary interval on $\mathbb{R}$ and let $L(I)$ be the set of all Lebesgue integrable functions defined on I.

Proposition 4.2 *If* $f \in L(I)$, *then there exists a sequence of radical functions* $(C_n)_n$ *such that* $\lim_{n \to \infty} C_n = f$ *at almost every point of* I.

Proof 4.3 *The proof takes into account that the countable union of null sets is a null set and that if* $f \in L(I)$, *then* $f = u - v$ *with* $u, v \in U(I)$, *where* $U(I)$ *is the set of all upper functions defined on* I. *In this case, there are sequences of step functions* $(s_n)_n$ *and* $(t_n)_n$ *such that* $u = \lim_{n \to \infty} s_n$ *at almost every point of* I *and* $v = \lim_{n \to \infty} t_n$ *at almost every point of* I. *So* $f = u - v = \lim_{n \to \infty} (s_n - t_n)$ *at almost every point of* I. *For more details, we refer to [Co3].*

Let I be a set defined by $I = I_1 \cup I_2$ where I_1 and I_2 are intervals such that $I_1 \cap I_2 = \emptyset$. Suppose that $f_1 \in L(I_1)$ and $f_2 \in L(I_2)$. Then, it is well-known that the function f defined by requiring to have value $f_1(x)$ at each point in I_1 and to have value $f_2(x)$ at each point in I_2 is a function that belongs to $L(I)$ and $\int_I f = \int_{I_1} f_1 + \int_{I_2} f_2$. So, the radical functions $C_{m,n}$ defined as (1.5) can approach this function $f \in L(I)$.

4.5 Approximation on $L^p[a, b]$ and $L^p(\mathbb{R})$, $1 \le p < \infty$

Let $([a, b], \mathcal{A}, \mu)$ be a measure space, where $\mathcal{A}$ is the σ- algebra of Borel subsets of $[a, b]$ and μ is the restriction of Lebesgue measure to $\mathcal{A}$. Let $\mathcal{L}^p([a, b], \mathcal{A}, \mu)$ be the set of all $\mathcal{A}$-measurable functions $f : [a, b] \to \mathbb{R}$ such that $|f|^p$ is integrable. The function

$\|\cdot\|_p : \mathcal{L}^p\left([a,b],\mathcal{A},\mu\right) \to \mathbb{R}$ defined by

$$\|f\|_p = \left(\int_a^b |f|^p \, d\mu\right)^{\frac{1}{p}}$$

is a seminorm. Let $L^p\left([a,b],\mathcal{A},\mu\right)$ be the set formed by identifying functions in $\mathcal{L}^p\left([a,b],\mathcal{A},\mu\right)$ that agree almost everywhere.

We shall use $\mathcal{L}^p\left[a,b\right]$ and $L^p\left[a,b\right]$ as abbreviations for the sets $\mathcal{L}^p\left([a,b],\mathcal{A},\mu\right)$ and $L^p\left([a,b],\mathcal{A},\mu\right)$ respectively.

Proposition 4.3 *The radical functions of the type (1.5) defined on $[a,b]$ are dense in $L^p\left[a,b\right]$.*

Proof 4.4 *Of course, each continuous function defined on $[a,b]$ as (1.5) belongs to $L^p\left[a,b\right]$. It is enough to prove that for each step function f and each positive number ϵ, there is a radical continuous function $C_{m,n}$ of the type (1.5) defined on $[a,b]$ that satisfies $\|f - C_{m,n}\|_p < \epsilon$. So, let f be a step function on $[a,b]$ and define, for each $n \in \mathbb{N}$, the radical continuous function $C_{m,n}$ by the formula (1.5) at each point x on $[a,b]$. In [Co3] one can see that for $n \geq n_0$*

$$\int_a^b |f - C_{m,n}| \, d\mu \leq \epsilon$$

for any $\epsilon > 0$

Let us call a function on $\mathbb{R}$ a step function if for each interval $[a,b]$ its restriction to $[a,b]$ is a step function. Analogous of Proposition (4.3) holds for $L^p\left(\mathbb{R},\mathcal{A},\mu\right)$ where $\mathcal{A}$ and is the σ- algebra of Borel subsets of $\mathbb{R}$ and μ is the Lebesgue measure on $\mathbb{R}$ if we replace the set of step functions on $[a,b]$ by the set of step functions on $\mathbb{R}$ that vanish outside some bounded interval, and if we replace the set of radical continuous functions on $[a,b]$ by the set of radical continuous functions on $\mathbb{R}$ that vanish outside some bounded interval.

4.6   Approximation to a measurable function

The function $f = 1$ is the limit of step functions on $\mathbb{R}$, however f is not a Lebesgue integrable function, $f \notin L(\mathbb{R})$. So, the set of functions that are limit of step functions, namely $\mathcal{M}(\mathbb{R})$, contains the set of Lebesgue integrable functions $L(\mathbb{R})$.

Theorem 4.1 *Let I be an arbitrary interval on $\mathbb{R}$ and let $f \in \mathcal{M}(I)$. Then there exists a sequence of radical continuous functions $(C_n)_n$ defined as (1.1) such that $\lim_{n \to \infty} C_n(x) = f(x)$ at almost every point of I.*

Proof 4.5 *The proof takes into account that the countable union of null sets is a null set and that if $f \in \mathcal{M}(I)$, then there exists a sequence of step functions $(s_n)_n$ on I such that $\lim_{n \to \infty} s_n(x) = f(x)$ at almost every point of I. If necessary, there is a complete proof in [Co3].*

Let f be a function defined on an arbitrary interval I and suppose that $(f_n)_n$ is a sequence of measurable functions on I such that $\lim_{n \to \infty} f_n(x) = f(x)$ at almost every point of I. Then, it is well-known that f is measurable on I. There are no measurable functions but this result shows us that it is not easy to construct examples of them. So, by means of the RAFU method on approximation we can approach almost all functions defined on an arbitrary interval I.

Bibliography

[Beh]   BEHFOROOZ, H., Approximation by integro cubic splines. Appl. Math. Comput. 175 (2006) 8-15.

[Be2]   BEHFOROOZ, H., Interpolation by integro quintic splines, Appl. Math. Comput. 216 (2010) 364-367.

[Bla]   BLASCO, J. L. and MOLTÓ, A. On the uniform closure of a linear space of bounded real-valued functions. Annali di Matematica Pura ed Applicata IV, vol. CXXXIV (1983) 233-239.

[Cas]   CASTILLO, E., IGLESIAS, A., GUTIÉRREZ, J. M., ÁLVAREZ, E., COBO, A. Mathematica. Editorial Paraninfo. 1995.

[Che]   CHEN, Y., HUANG Y., HAN, W., Function reconstruction from noisy local averages, Inverse Problems 24 (2008) 025003. http://dx.doi.org./10.1088/0266-5611/24/2/025003. p.14

[Co1]   CORBACHO, E. Uniform approximation with radical functions, $S\overrightarrow{e}MA$ Journal 58 (2012) 97-122.

[Co2]   CORBACHO, E. Corbacho, A RAFU linear space uniformly dense in $C[a,b]$, Appl. Gen. Topology 14 (1) (2013) 53-60.

[Co3]   CORBACHO, E. Approximation in different smoothness spaces with the RAFU method, Appl. Gen. Topology 15 (2) (2014) 221-228.

[Del] DELHEZ, E., A spline interpolation technique that preserves mass budgets, Appl. Math. Lett. 16 (2003), 16-26

[Eps] EPSTEIN, E., On obtaining daily climatological values from monthly means, J. Clim. 4 (1991) 365-8

[Ess] ESSEN, C. G. Über die asymptotisch beste Approximation stetiger Funktionen mit Hilfe von Bernstein-Polynomen, Number. Mat. 2 (1960), 206-213

[Mon] GARRIDO, M. I. and MONTALVO, F. Uniform approximation theorems for real-valued continuous functions. Topology Appl. 45 (1992), 145-155.

[Gas] GASSÓ, T., HERNÁNDEZ S. and ROJAS, E. Representation and approximation by series of continuous functions. Acta Math. Hungar. 133 no. 1-2 (2009) 91-102.

[Gon] GONZÁLEZ, R. C., WOODS, R. E., Digital image Processing 2nd edn, New Jersey: Prentice Hall, 2002

[Hua] HUANG, J., CHEN, Y., A regularization method for the function reconstruction from approximate average fluxes, Inverse Problems 21 (2005) 1667-84.

[Jam] JAMESON, G. J. O., Topology and normed spaces., Chapman and Hall, (London, 1974).

[Kak] KAKUTANI, S., Concrete representation of abstract (M)-spaces, Ann. Math., 42 (1941), pp 994-1024.

[Kil] KILLWORTH, P., Time interpolation of forcing fields in ocean models, TJ. Phys. Oceanogr. 26 (1996) 136-43

[Lor] LORENTZ, G.G. Bernstein polynomials. Chelsea Publishing Company. New York. 1986

[Mro] MROWKA, S., On some approximation theorems. Nieuw Archief voor Wiskunde (3) XVI (1968) 94-111.

[Ram] RAMÍREZ GONZÁLEZ, V., BARRERA ROSILLO, D., PASADAS FER-
       NÁNDEZ, M., GONZÁLEZ RODELAS, P. Cálculo Numérico con Mathemat-
       ica. Ariel Ciencia. 2001

[Si1] SIKKEMA, P.C., Über den Grad der Approximation mit Bernstein-
       Polynomen, Number. Mat. 1 (1959), 221-239

[Si2] SIKKEMA, P.C., Der Wert einiger Konstanten in der theorie der Approxima-
       tion mit Bernstein-Polynomen, Number. Mat. 3 (1961), 107-116.

[Sto] STONE, M.H., Applications of the theory of Boolean rings to general topology.
       Trans. Amer. Math. Soc., 41 (1937), pp. 375-481.

[St2] STONE, S., A generalized Weierstrass approximation theorem. Math. Maga-
       zine, 21 (1948) 167-184, 237-254.

[Tie] TIETZE, H., Über functionen die anf einer abgeschlossenen menge stetig sind.
       Journ. Math., 145 (1915) 9-14.

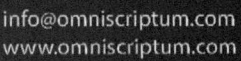